Bee Nilson

Mrs A. R. (Bee) Nilson, B.H.Sc., S.R.D.
Dip. Ed., M.I.M.A., M.R.S.H., was born
in New Zealand, where she received her
professional training. For 30 years she
has lived and worked in Britain, putting
her training into practice in running
her own home and in teaching and
writing about food and nutrition. Among
her publications are *Pears Family
Cookbook*, *The Penguin Cookery Book*,
Cooking for Special Diets and *The Book
of Meat Cookery*. For many years she
was Senior Lecturer in nutrition at the
Northern Polytechnic, London.

Deep Freeze Cooking

Bee Nilson

A Mayflower Handbook

DEEP FREEZE COOKING

Bee Nilson

Copyright © Bee Nilson 1969

First published in Great Britain by
Pelham Books Ltd. 1969
Published as a Mayflower Handbook 1970

Mayflower Paperbacks are published by
Mayflower Books,
3 Upper James St., London, W.1.
Made and printed in Great Britain by
C. Nicholls & Company Ltd.,
The Philips Park Press, Manchester

Contents

INTRODUCTION

When I started to use a freezer several years ago, I wanted to know many things about it. How it worked, how to freeze the raw materials of cooking, how to freeze everyday recipes and special-occasion dishes, which of my recipes were suitable for freezing and which needed adjusting.

As I became more experienced I learnt that very few recipes needed altering in any way but the information I did need for easy reference was the best way to freeze foods of different kinds, how to thaw and heat them, how long it takes to cook things straight from the freezer, which were better cooked this way and which needed to be thawed first.

Because I am sure this is what other people want to know about a freezer, the bulk of this book is concerned with how to freeze foods and how to use frozen food. In addition, I have included a selection of recipes for dishes I freeze. Where only certain methods give good results in the freezer, for example, with making ice-cream, I have included several recipes. Others are intended chiefly as an indication of the type of recipe I find it useful to freeze.

WEIGHTS AND MEASURES

ALL MEASURES USED IN THIS BOOK ARE LEVEL MEASURES

Both weights and measures are given in the recipes because some prefer to weigh when cooking and others like measuring. Either method gives good results provided reasonable care is taken. The measures used in the recipes are British Standards Institute Kitchen measures. They have the following capacities:

1 cup (c) $= \frac{1}{2}$ Imperial pint
$\quad\quad\quad = 284$ millilitres (ml)
$\quad\quad\quad = 10$ fluid ounces
1 tablespoon (Tbs) $= \frac{1}{32}$ of an Imperial pint
$\quad\quad\quad\quad\quad = 17 \cdot 7$ ml
1 teaspoon (tsp) $= \frac{1}{96}$ of an Imperial pint
$\quad\quad\quad\quad = 5 \cdot 9$ ml

3 tsp $= 1$ Tbs
16 Tbs $= 1$ cup
16 Tbs $= \frac{1}{2}$ pint

METRIC SYSTEM

In all recipes British weights, and many of the measures, have been converted to the metric system. This has not been done with tablespoons and teaspoons but sizes can be checked against the capacities given in the list above.

The conversion has been adjusted to give practical metric weights and measures which are still sufficiently accurate to give good results. The following figures have been used:

1 oz $= 30$ grammes (g)
1 lb $= \frac{1}{2}$ kilogramme (kg)
2 lb $= 1$ kilogramme (kg)
$\frac{1}{4}$ pint $= 1\frac{1}{2}$ decilitres (dl)
$\frac{1}{2}$ pint $= \frac{1}{4}$ litre (l)
1 pint $= \frac{1}{2}$ litre (l)
$\frac{1}{8}$ inch $= 3$ millimetres (mm)
1 inch $= 2\frac{1}{2}$ centimetres (cm)

TEMPERATURES

In the general information on freezing, temperatures are given in centigrade followed by the equivalent fahrenheit number.
Oven temperatures in the recipes are for electric cookers. Gas mark numbers are also given.

The following is a useful table for those who want to convert from fahrenheit to centigrade.

	°F	°C
Storage in the freezer	0	−18
Storage in the refrigerator	40–50	5–10
Freezing temperature of water	32	0
Simmering	185	85
Boiling	212	100
Tepid or lukewarm	80	30

Oven Temperatures	°F	°C
Very Slow	250–300	120–150
Slow	300–350	150–177
Moderate	350–375	177–190
Moderately Hot	375–400	190–205
Hot	400–425	205–218
Very Hot	450–500	232–316

Chapter One

THE DEVELOPMENT OF DEEP FREEZING

It seems that people living in the Arctic were the first to use the technique of freezing as a method of food preservation. Living at temperatures well below zero, they found that food left outside froze quickly and kept indefinitely.

Clarence Birdseye is credited with being the first visitor to the Arctic who made use of this technique for preserving foods in temperate climates. In 1928 he went on a hunting and fishing holiday in Labrador and noted how the Esquimaux preserved fish and caribou meat by freezing it quickly in the open air. Having found the food to be tender and fresh-tasting even after months of storage, he proceeded to experiment and found the secret was the rapid rate of freezing, only possible at very low temperatures. Clarence Birdseye sold his invention to General Foods Corporation and the first commercial products were marketed in the United States of America in 1929.

By 1937 up to 500,000 hundredweight of fruit and 150,000 hundredweight of vegetables were being frozen and by 1945 the total production was 400,000 tons. In the early days the products were marketed almost entirely in large containers, the small household packs coming later.

In the United Kingdom commercial deep freezing did not start until after the Second World War. In 1946 there were only 100 retailers selling frozen food, but by 1967 there were over 100,000.

Figures for 1968 show remarkable increases in the volume of frozen foods used both in catering and by the housewife. Some of the frozen food is produced in the United Kingdom but much of it comes from abroad. For example, New Zealand, which since 1882 has been sending food to Britain using methods of cold storage, is today also sending quick-frozen foods such as peas.

Research and development are continuous and undoubtedly the next few years will see not only an increase in the volume of frozen foods used, but also improvements in quality. Some of the interesting developments that have already taken place are in methods of commercial freezing.

Early methods were by what is known as plate freezing. The plates are metal surfaces which are at a very low temperature. The food is placed on narrow shelves and the freezing plates are then brought into close contact with it. The plates conduct heat away from the food. Food frozen in this way is usually in block form. It is a quick and efficient method, still in use for many foods.

Then came the development of air-blast freezing where cold air passes over the food and freezes it. This is either carried out with small amounts of food in "Batch Freezing", or with large amounts of food on trucks passing through a tunnel of cold air, known as "Tunnel Freezing". These are good methods for producing individually frozen pieces of food rather than the solid blocks.

One of the latest developments is the use of liquid nitrogen. Nitrogen gas is harmless to humans. It forms four-fifths of normal air and is continually entering and leaving the lungs. It does not react with foods and so cannot produce harmful changes. When a gas is compressed it becomes liquid. The temperature of liquid nitrogen is $-196°C$ ($-320°F$) and food can be quickly frozen by either immersion in the liquid or by spraying, the latter technique being quicker and more effective. Results show that the taste and texture of foods such as strawberries, asparagus and many others are retained better, even after thawing, than with older methods of freezing.

The method of freezing in most domestic freezers is on the same principle as "Plate Freezing". The food to be frozen is either in contact with freezing shelves or the sides and floor of the cabinet. Some larger models incorporate a fan which blows cold air round the food being frozen, but the air is not at such a low temperature as that used in the factory plant.

The actual location of the commercial freezing plant is changing too. All fish frozen in the United Kingdom used to be brought to port packed in ice and then frozen in factories near the port. Now, in addition, there are trawlers which will freeze whole gutted fish and thus bring it to port in better condition than when it is merely stored in ice. In addition, there are a few freezer factory ships where the fish is cut up and frozen at sea. The factory ship is supplied by a fleet of trawlers.

Plants for freezing fruit and vegetables have always been located near the farms in order to get the produce from field to factory in the shortest possible time. Now some farmers are doing their own freezing in small units on the farm. Small plants using liquid nitrogen are being developed for this purpose and it has special advantages with crops such as strawberries and peas.

Frozen prepared meals have been in use in the United Kingdom for more than twenty years. They were first developed for use on airliners. Frozen meals are today available in a limited range for domestic use but are being more widely used by the caterer. Many caterers are convinced that this is the answer to many problems in hospitals and other large institutions. Some catering establishments purchase commercially frozen foods while others use their kitchens as a factory to produce and freeze their own dishes. Thus the kitchen can work steadily throughout the day producing food for meals for many days ahead.

Methods of packaging the food are another subject of research and improvement. This is a very vital factor in preserving the quality of the frozen food during storage and marketing. It also affects the cost to the consumer.

With the coming of the domestic freezer the housewife is not only enabled to keep supplies of commercially frozen products at hand, but also to freeze her own surplus garden or farm produce. The freezer also becomes a convenient tool to save time and labour in meal preparation and shopping. Our mothers and grandmothers had baking days when bread, pastry and cakes would be made for a week. Deep freezing enables our generation to extend the idea of a baking day to cover all types of food. This cooking can be done at a time convenient to us and the results used to provide meals for weeks ahead. It has obvious advantages for entertaining when the hard work can be done days ahead. I find, too, that I enjoy my own cooking more when the results are served at a decent interval after the labour of producing them. There is something rather special about serving and eating a good home-cooked meal which has involved nothing more strenuous in immediate preparation than thawing and heating the food and preparing the trimmings.

Chapter Two

FACTS ABOUT DEEP FREEZING

Freezing is low-temperature food preservation. Of all the methods man has devised for preserving food deep freezing produces a product nearest to the fresh in quality. Modern man with his urban society, increasing birth rate and longer expectation of life is as dependent on preserved foods for survival as were many of his ancestors who devised means of keeping food from one harvest to the next. When they were not successful in achieving this, hunger and death often resulted.

Deep freezing is the term applied to domestic freezing, usually carried out at temperatures in the range of -23 to $-29°C$ (-10 to $-20°F$); quick freezing is the commercial method carried out at lower temperatures in the range of -29 to $-40°C$ (-20 to $-40°F$) or lower. The lower freezing temperatures of the commercial plant give faster freezing, which is an advantage in a factory, enabling more food to be frozen in a given time and thus reducing operating costs. In addition, the faster the freezing the better the quality of the frozen product. This advantage is sometimes lost by inadequate transport and storage facilities so that it is possible for food frozen in a domestic freezer at higher temperatures to be better in the end because there has been better control of the storage conditions.

The freezing temperature of water is $0°C$ ($32°F$). Most foods will freeze at this temperature but take longer to become frozen, and when they are thawed there is a marked loss of quality, with collapse of watery foods like fruit and vegetables and loss of juice or "drip" from animal tissues. This is because water has been withdrawn from the cells during the slow freezing. With quick or deep freezing, on the other hand, minute ice crystals are formed within the cells. The result is that the thawed cell is not a dehydrated cell and the thawed product tastes and looks like a fresh one.

When storage conditions are bad, that is the temperature too high or allowed to fluctuate, water is withdrawn from the cells

14

and the size of ice crystals outside the cells increases, leading to tissue damage and loss of liquid on thawing.

Storage temperatures are, therefore, as important as freezing temperatures in producing good-quality frozen foods. This is where home freezing has an advantage over commercial freezing in some cases. The housewife has control over the storage conditions of food she has frozen herself whereas she has no idea what happens to the food between the freezer factory and her kitchen until she opens the packet, possibly to find a poor-quality product inside. Sometimes, of course, this is her fault. She may have purchased a good frozen article but let it thaw on the way home and then re-frozen it.

WHY FREEZING PRESERVES FOOD

Freezing slows down chemical action and the activity of bacteria and fungi, all of which cause food spoilage. It inactivates enzymes which bring about the ripening of fruits and vegetables, the "ageing" of meat, and finally food decay. The activity of micro-organisms which cause decay is also stopped. Oxidation which causes rancidity of fats and discoloration of fruits and vegetables is reduced to a minimum provided the food is correctly packed. With long storage, however, oxidation does slowly take place which is why certain foods such as fatty meats and oily fish and some fruits have a comparatively short storage life if they are to be of good quality when consumed.

FACTORS AFFECTING QUALITY IN FROZEN FOODS

The most important of these are the quality of the food before freezing; the temperature used and therefore the speed of freezing; the storage temperature and length of storage; the method of packing; and the method of thawing.

To ensure speedy freezing in the domestic freezer it is important to follow the recommendations of the manufacturer. This is particularly important in regard to the amount of food which can safely be frozen in 24 hours and the best position in which to put it for most rapid freezing.

Storage temperature has an even greater effect on quality of product than freezing temperature, for reasons already given above. The process of deterioration is hastened by temperatures above $-18°C$ ($0°F$) and even more by fluctuating temperatures of more than $4°C$ ($5°F$) above or below this. That is why tables of recommended storage times for different foods are usually given for a specific storage temperature such as $-18°C$ ($0°F$). During transport and sale of commercially frozen foods careless handling can cause wide fluctuations of temperature with possible deterioration of colour, flavour and appearance. This sort of thing can happen in a domestic freezer if too much food is put in for freezing at a time or if the food is warm or is put in contact with already frozen food. Excessive frost inside a package is an indication that this has happened and it is usually worse if there is a lot of air left in the package.

Good packing protects the food against air and the effects of oxygen which causes spoilage of fats and colour changes. Freezing in a syrup (fruit) or a sauce or stock (meat and fish) helps to exclude oxygen. Moisture/vapour-proof packaging helps to prevent evaporation of moisture with consequent dehydration and other undesirable changes in the food.

In general, the more rapidly thawing is carried out the better the final product. This is particularly the case with vegetables, where damage to the cells is much less if the frozen product is put into boiling water and cooked for the minimum time. Fillets and slices of meat and fish are usually better if cooked frozen, or at least before they have completely thawed. When large items like joints of meat and poultry are to be thawed before cooking it is wiser to do this in the refrigerator or cold larder. There are two reasons for this: first there is less loss of liquid and less damage to the tissues; secondly, if the outside thaws too quickly, as it does at room temperature, spoilage can start before the inside is thawed. To prevent contamination of the outside the joint or bird should be thawed in its original wrapping.

Chapter Three

HEALTH ASPECTS OF FROZEN FOOD

Deep freezing not only retains the natural flavour of foods better than any other method of preserving, but also causes the least loss of nutritive value. Losses due to the actual freezing process are minimal but small losses do occur during preparation, storage and thawing. The extent of these losses depends on the methods of preparation and thawing used and on storage conditions. The extent of such losses varies with different foods. As is the case with using fresh foods, the nutritive value of frozen foods can be reduced by poor cooking and serving methods.

MEAT AND FISH

The nutritive value is unaffected by the freezing process. Experiments have shown that there are losses during storage; for example, pork chops stored for a year at −18°C (0°F) lost about 10 per cent of the thiamine and riboflavine but the niacin was unaffected. It will be appreciated that there are nutritional advantages in having a quick turnover of frozen food rather than aiming at very long storage times. This applies to other foods and other nutrients too.

When meat or fish is thawed there is some loss of juice and with the juice some vitamins, so that it is important to keep "drip" losses to the minimum. This is achieved by thawing in the refrigerator and cooking as soon as the meat or fish has thawed, or even before it is completely thawed.

VEGETABLES

Prior to freezing, vegetables undergo a blanching process and this causes loss of some nutrients (vitamins and minerals), but it also serves to inactivate enzymes which themselves bring about destruction of vitamin C. There are therefore both gains and losses due to blanching. An example of the extent of the loss can be seen with peas. Peas blanched for less than 1 minute lost 25 per cent of the vitamin C and 12 per cent of the minerals. Similar results have been noted for the vitamins riboflavine and niacin. There can be 0–20 per cent loss of carotene during blanching. There will be further losses during the cooking process. The total losses are very much the same as with preparing and cooking fresh vegetables and it is possible to have a higher nutritive value in frozen vegetables than in purchased fresh vegetables just because the frozen ones were fresher at the beginning. For example, fresh vegetables kept for a day or two after gathering and before cooking can lose 50 per cent of the vitamin C. Frozen ones are processed very soon after being gathered and the losses are less.

When frozen vegetables are purchased and stored in a refrigerator, and not a freezer, there is still good retention for a day or so provided the packet is not opened and air allowed to get in.

Because vegetables have been blanched before freezing cooking times are less than with fresh vegetables and this offsets the losses due to blanching before freezing. It is important to blanch by the quickest possible method and to cool the vegetables quickly afterwards (see page 119 for details).

During cold storage at $-18°C$ (0°F) and below no significant vitamin losses occur. With higher storage temperatures, e.g. $-12°C$ (10°F), levels of vitamin C may fall by 25 per cent. To be fair to frozen vegetables, however, it is necessary to point out that even these losses are less than those commonly found with vegetables sold through shops and markets. Only home-grown and quickly used vegetables have a higher nutritive value than the frozen ones.

Proper cooking methods are important in preventing losses during thawing (see pages 121–9).

FRUIT

Vitamin C is the most important nutrient obtained from fruit. About 20 per cent of this vitamin is lost during the combined processes of freezing and thawing. An unnecessarily long thawing period, especially at room temperature, can increase the losses. Fruit frozen in a sugar syrup is less likely to suffer as much as fruit frozen dry. Both flavour and nutritive value are at their best when the fruit is eaten barely thawed or when it is cooked while still frozen.

FAT

Either fat alone or fat in other foods can oxidise and become rancid, especially if the food is stored for a long time. Oxidation also causes losses of fat-soluble vitamins A, D, E, and K.

CAN FROZEN FOOD BECOME UNSAFE TO EAT?

No food-poisoning bacteria have been found able to grow at temperatures even as high as $4.5°C$ (40°F). Temperatures of -10

to 4·5°C (14 to 40°F) permit slow growth of some other organisms and slow spoilage of the food without danger to health.

But although food-poisoning organisms do not grow at these low temperatures they do survive. The survival rate of micro-organisms is around 50 per cent, the rate continuing to decrease with longer storage time. There is danger of food poisoning from frozen foods only if they are mishandled before packing or during distribution, thawing, cooking and serving.

Certain parasites, such as those in pork which cause trichinosis, are destroyed at freezing temperatures.

The aim in preparing foods for deep freezing should be to keep them as clean as possible. This applies particularly to pre-cooked foods. Dirty handling of these foods can raise the micro-organism population considerably.

Re-freezing of partly thawed food is not harmful to health provided the packet has not been opened, but the quality of the food deteriorates. Keeping fully thawed foods such as meat, fish, eggs and prepared foods at temperatures above those in a normal domestic refrigerator for any length of time can be dangerous; how dangerous will depend on the degree of infection in the food before freezing.

In general, the sort of clean handling which is good policy with fresh food is good policy with frozen food. Mishandling of either can cause food poisoning.

Chapter Four

CHOOSING A FREEZER

Apart from the question of cost, the most important factor for most people is the space available in which to put the freezer. Even within the limits of these two factors there are some other points worth consideration as well.

An important point to which not everyone pays sufficient attention is what the freezer will be used for. For example, will you be wanting to freeze your own garden or farm produce? Then you will need a freezer large enough to store the food for several months, as well as to store foods for current use.

Do you intend to do bulk buying of commercially frozen foods? You will want a larger freezer than the person who intends to carry only a small stock of these foods. If you intend to use the freezer to enable you to reduce bulk shopping to once a month then you will obviously need a larger freezer than the person who is happy to shop weekly.

Do you entertain a lot and will you want a good stock of pre-cooked and pre-prepared dishes to use for unexpected guests? Some of the more elaborate foods can be quite bulky to store. Will you want to prepare food for large parties days or even weeks ahead? Or is there someone in your household who goes shooting or fishing and brings home food to be stored?

Is yours a family that likes cakes, pastries and other baked goods and do you hope to use the freezer to store these and cut baking days to one a month? Perhaps you are often away from home for several days or weeks and you want to leave a supply of ready-prepared food for the family. If you have a full-time job outside the home you will probably want to use the freezer in this way too.

Does your family love fresh bread and is it impossible to get it every day? In that case you will want to have room in the freezer to store bread and rolls. Or are you a small family where bread tends to go stale before you are half-way through a loaf? Then you will want room to keep half loaves, sliced bread and rolls.

And what about puddings? Will you want to bulk-buy commercial ice-cream? Will you want to carry a good supply of pastry, pies and puddings?

If members of the family have strong likes and dislikes you will want to use the freezer to enable you to satisfy varied tastes without trouble. If there is someone on a special diet you will want to cook several portions of a dish and freeze the surplus.

When there are children in the family you will want to have a large enough freezer to enable you to do extra cooking in anticipation of holidays, to be free yourself.

It is worth while giving some thought to these factors, remembering that so many people find they have bought too small a freezer for their needs because they did not realise all the uses to which it could be put.

TYPES OF FREEZERS ON THE MARKET

Most manufacturers bring out new models every year. It is thus impossible to give up-to-date technical details. The most useful thing is to consider the pros and cons of the two basic shapes, the chest type which is a box with a lid, or the upright front-opening type which looks like an ordinary refrigerator.

Chest Type The chief advantage of this type is that there is practically no loss of cold when the lid is opened. This is because cold air is heavy and tends to fall.

The chest type has certain disadvantages if you want to do your own freezing. For good freezing, food should be in contact with the sides or bottom of the chest and it is better for already frozen food if it is not in contact with food being frozen. To arrange this can be quite a problem unless the freezer has a special compartment for freezing, which some large models can offer.

A further point to consider is that if you are a short person you should check the model before buying to make sure you can reach the bottom back portions easily.

Chest-type freezers take up more floor space than the upright ones, and while in theory the top can be used for an extra table top, in practice it is a nuisance to have to lift things off to open the freezer.

Some chest models are as small as 3–4 cu ft (84–112 litres) but the majority of domestic models are between 6 and 12 cu ft (168–336 litres).

Front-opening type or Upright model The chief disadvantage of this type is that cold air is lost every time the door is opened. But there are many compensating advantages.

It is the better type to have if you want to do much freezing of your own products. A good freezer of this type should have several of the shelves with freezer coils in them so that the food freezes quickly. Check this point because some models have only storage shelves and offer no advantage over the chest type in this respect. Some upright models have plastic drawers to hold the food while others have wire baskets. Check these to make sure they will move easily when they are full of food.

The upright type takes up less floor space than the chest models but is taller. Be sure there is enough free height to take it and that you can reach the back of the top shelf.

If you want only a small freezer those of table-top height can be built in and give an extra working top.

It is important to check that the floor is strong enough to take the weight of a large freezer full of food, especially the upright model standing on a small portion of the floor. You may also need to check that there is room to open the door fully to allow racks to be withdrawn.

The smallest models of the upright type are 1–2 cu ft (28–56 litres) and are made to stand on top of a domestic refrigerator or any working surface. Others are from 4 cu ft (112 litres) upwards to the large commercial models. Some are double-deckers, with a freezer at the bottom quite separate from the refrigerator above.

The freezer compartments in some refrigerators are for storing frozen food for short periods only and are not suitable for freezing and storing large amounts of food.

WHAT SIZE TO BUY

This will depend on a number of factors in addition to the size of the family (see page 19).

Each cubic foot (28 litres) of freezer space will hold approximately 25–30 lb (14–15 kg), depending on the shape of the packages, and will freeze about $2\frac{1}{2}$ lb (1 kg) of fresh food at a time. If the freezer is to be used for bulk storage of produce allow 6 cu ft (168 litres) of freezer space per person. If it is only for storage for current needs 3–5 cu ft (84–140 litres) of freezer space per person will be sufficient.

COSTS

Running costs depend on the size of the freezer; the way it is made; the number of times the lid or door is opened per day and for how long; the temperature of the food when it is put in (advisable to cool it in the refrigerator first); and local electricity charges.

The following figures are a guide to estimating the probable running costs:

Approximately 6 cu ft (168 litres) freezer consumes

		1·8 Kw in 24 hrs.
12 cu ft (336 litres)	...	3·0 Kw in 24 hrs.
18 cu ft (504 litres)	...	3·6 Kw in 24 hrs.

These figures assume that the freezer motor will be running for 12 hours out of the 24. When food is being frozen the consumption will be higher.

To estimate depreciation costs take the average life of a freezer to be about 15 years.

Whether owning a freezer proves to be an economy depends on many factors but there is no doubt about its importance as a convenience to the housewife and caterer. The faster the turnover of food the less it costs per pound for freezing and storage. Remember too that the operating costs are just the same whether the freezer is full or half full.

Small economies in cooking fuel can be made by cooking more than enough for one meal and freezing the rest for short periods. This particularly applies to food which will simply need to be thawed before serving. For those requiring re-heating it can take as much fuel as to cook it in the first place.

FEATURES TO LOOK FOR

A good freezer must maintain an average temperature not higher than $-18°C$ ($0°F$), preferably below this. If you want to do much freezing yourself it should be adjustable to go down to $-29°C$ ($-20°F$) for the freezing process and should be able to freeze the food to $-18°C$ ($0°F$) within 24 hours. It should be able to do this for up to one-tenth of its capacity in the 24 hours. Fluctuations in temperatures should not be more than plus or minus $4°C$ ($5°F$), preferably less. Wide fluctuations in temperature during storage mean loss in quality.

The size of the compressor should be such that the motor runs for only 12 out of the 24 hours when food is simply being stored. It may run very much more during the actual freezing process, or if the freezer has been placed in a bad position.

Make sure the door has a good seal to retain the cold and prevent excessive formation of frost in the cabinet. Examine defrosting arrangements to see what happens to the melted ice. Some models have automatic de-frosting.

WHERE TO PLACE IT

Avoid a damp place as this can damage both the motor and the exterior of the cabinet. Do not put it in a place that is likely to go below 4·5°C (40°F) in cold weather, unless the freezer is especially designed for this purpose.

In or near the kitchen is the most convenient place to have it, but it can be put in a dry garage, larder or basement. It should not be fitted tightly into any space but needs good ventilation all round.

Test to see that it is standing level or the door may warp, spoiling the fit. Test the door fit by inserting a piece of paper. Close the door on the paper; then it should not be possible to pull the paper out without tearing it.

Avoid a hot room or a position in full sun or by a radiator. If it is in a hot place the motor will have to run more than it should to keep the temperature down.

Chapter Five

PACKING MATERIALS AND METHODS

The right choice of materials and methods used for packing frozen food is essential for successful results. The longer the food is to be stored the more important this becomes. As the cost of the best packing materials is only a fraction of the cost of the food it is worth while having the best. Materials suitable for use in the freezer are also suitable for the refrigerator but the reverse is not always true. However, for short-term storage of 3–4 weeks materials such as the waxed paper in which sliced bread is sold, or single household foil are suitable.

There are a number of reasons why it is worth while taking care with packing. Air in the freezer is dry and will draw moisture from the food unless the latter is protected by a moisture/vapour-proof cover. The longer the food is stored the greater the loss of moisture. Freezer "burn", to which meat, poultry and fish are particularly susceptible, is caused by excessive drying on the surface, made worse hy fluctuating storage temperatures. Freezer burn shows up as amber patches on meat and poultry and white opaque patches on fish.

As well as protecting the food from air in the freezer, it is important to exclude as much air as possible from the package.

24

This is achieved by tight packing and elimination of air pockets before sealing. Air should be pressed out of bags and rigid containers should be of a size that will leave the minimum space when the food is frozen. For the allowance needed for expansion during freezing, see page 30.

By excluding as much air as possible the risk of oxidation of fat and the consequent rancidity is reduced. Oxygen also causes losses of vitamin C in fruits and vegetables.

Other reasons for having an airtight package are the need to preserve texture and flavour, to prevent excessive frosting in the freezing cabinet, and to keep the food clean.

Materials used must be able to stand up to the very low temperatures in the freezer, should be odourless, non-absorbent to fat and blood, and non-toxic. Bags and wrapping material should be pliable enough to mould close to the food and to seal satisfactorily. If rigid containers are used, rectangular ones are better than round or cylindrical because they pack together without wasting freezer space.

TYPES OF MATERIAL AVAILABLE

These can be broadly divided into sheet materials for wrapping, bags, and rigid containers (boxes, tubs, etc.). The firm supplying your freezer may keep a stock of these but, if not, should be able to tell you where you can get them locally. Large stores with a good household stationery department usually carry some stocks and specialist firms supply by mail order, for example, Frigicold Ltd, Manchester Square, London, W.1, or Lawson's Ltd, 1A St Andrew's Street (South), Bury St Edmunds, Suffolk.

The following information covers materials fairly readily available at the time of writing. Some materials (not listed) are only available for commercial use, but may eventually be offered for domestic use.

SHEET WRAPPING MATERIALS

Aluminium Kitchen Foil This is very useful and satisfactory provided it is used with care. It has the great advantage that it can be moulded closely to the food, thus excluding air. It is not suitable for use in single thicknesses except for very short storage periods. This is because it very easily becomes a mass of small holes, hence an inadequate protection. It is a wise precaution not only to use it double but also to wrap very generously. It if has an overwrap of mutton cloth, plastic bag, nylon bag or other protection this will help prevent damage to the foil when packages are moved about in the freezer. For very long storage it is advisable to overwrap with freezer paper.

A further advantage of foil is that food can be put in the oven to heat, still in the wrapping. It is also useful for lining casseroles, pie plates and other cooking equipment (see pages 29, 98 and 181).

I find foil is inclined to stick to meat and fish if they are unwrapped while still frozen, but dipping the package in warm water before opening it loosens the foil.

KVP Freezer Paper This is a very strong moisture/vapour-proof paper which has several advantages. It has a special coating inside and an uncoated surface outside on which to write the label. It resists low temperatures without becoming brittle, is highly resistant to fats and the package can be easily sealed with freezer tape. The paper is sold in rolls.

Polythene Sheets These are useful for wrapping large articles and odd-shaped ones. It clings to the food better than a polythene bag does and can be cut to make small parcels.

Cellophane This must be a specially prepared heat-sealing Cellophane. Ordinary Cellophane becomes brittle at freezer temperatures. It is very useful for wrapping small articles separately within a larger pack and for separating layers of food—for example, chops, fillets, small cakes, etc. If it is used as the main wrapper it is wise to protect it with mutton cloth or other cloth or even brown paper.

"Look" Cooking Film This is a polyester film and can be used in the same way as polythene sheets.

POLYTHENE BAGS

It is important to use only the special-quality ones made for freezer use. The ordinary ones are not sufficiently moisture/vapour proof. Bags can be heat-sealed by putting a piece of paper between the polythene and a warm iron. Or seal by twisting the top into a bunch and securing with a piece of plastic-coated wire or a pipe cleaner. Bags are sold in a range of sizes from very small —4 inch (10 cm)—to very large for chickens, turkeys or salmon. The bags can be washed and re-used provided they have not become punctured during storage. Test for this by filling the bag with water while you are washing it.

With large bags of food it is wise to over-wrap the bag with either brown paper, greaseproof, clean nylon stockings or mutton cloth to prevent the bag from being torn when food is moved about in the freezer. Polythene bags are useful as liners for cartons for easy removal of the food. They can be made into rectangular packages by putting the bag in a carton to fill, freeze in the carton and then remove for storage.

The disadvantage of polythene bags is that they are not ideal for very long storage because they are not completely oxygen-proof.

RIGID CONTAINERS

It is advisable to use only plastic and polythene containers made specially for freezer use. Some other types may contain materials which produce toxic substances under freezing conditions.

Rigid Plastic Boxes Those for freezer use are made in a limited number of sizes. Although they have close-fitting lids it is advisable to seal the lids on with freezer tape (polytape).

Polythene Boxes Those specially made for freezer use are available in a variety of shapes and sizes. They have self-sealing lids and do not require tape.

Waxed Tubs These are wax-coated tubs with press-on lids. They are useful for small amounts of soft foods and liquids. With care they can be used more than once. Careless handling can crack the wax coating so it is a wise precaution to use a polythene bag as a liner the second time they are used.

They are not as durable as polythene boxes but are cheaper. They are sold in sizes from 8 fluid ounces ($2\frac{1}{2}$ dl) to 80 fluid ounces

($2\frac{1}{2}$ l) and spare lids are available, which is useful, because the lids sometimes get damaged during removal.

They can be stacked one above the other but have the disadvantage of being round and thus waste some space.

Screw-topped waxed boxes are also available in several sizes. They are stronger than the other type.

No waxed containers are suitable for filling while food is still hot—the wax coating will melt.

Waxed Cartons and Boxes These are sold with or without polythene liners. Those without liners need to have the tops sealed with polytape but the lined ones do not need sealing provided the liner has been properly sealed. The lids are either fitted or tuck-in.

Tins Those in which food has been purchased can be used provided they are rust-free. They are useful for the storage of fragile goods. It is best to use either a polythene liner or to wrap the foods individually in Cellophane paper. In this case it is not necessary to seal the tin with tape.

Aluminium Containers These are suitable if they have airtight lids or can be sealed with polytape. They are good for re-heating foods, e.g. steamed puddings. Do not use them for very acid or salty foods.

Casseroles It is wasteful to store these in the freezer because they take up a lot of room and are out of use for some time. Instead, line the casserole with foil before cooking the dish to be frozen, freeze in the casserole, remove the foil and food, wrap and store. The contents can then be returned to the original casserole for thawing and heating. An alternative method is to cook the food without a foil lining. Freeze it in the casserole and then dip the casserole in warm water to loosen the contents, remove and wrap.

If a very large amount of food has been cooked at a time it is better to divide it up into smaller containers for freezing.

Glass Jars These are suitable provided they have airtight lids. They should not be filled too full (leave 1–2 inches, or $2\frac{1}{2}$–5 cm) or they will burst. Unless they have straight sides the food will have to be completely thawed before it can be removed.

MISCELLANEOUS FREEZER MATERIALS

Aluminium Foil Dishes These include patty tins, round pie dishes, square-ended pie dishes, pie plates, compartment trays (for freezing one-portion meals), basins and steak-pie dishes. For freezing they should be over-wrapped with double foil or other freezer wrap. The food is thawed and re-heated in the container which can often be used more than once. They are a great convenience for pies, puddings and made-up dishes.

Waxed-paper Sweet Cups These are useful for freezing single eggs, individual portions of sweets or ices and savouries. They can be frozen unwrapped on a tray and then stacked in bags or boxes for storage.

Polytape This is a sealing tape specially made to withstand freezer temperatures which ordinary sealing tapes will not do. It is usually available in 1 inch (2½ cm) wide rolls. It is used for sealing the lids of boxes, the folds in parcels, and fastening on labels.

Plastic-coated Wire or Pipe Cleaners Unless you intend to heat-seal all polythene bags, you will need a stock of these wires to secure the tops of the bags.

Special Labels These are small white labels made to adhere to any surface and to stay on during freezing. They cannot be made to stick if put on after the food has been frozen. Ordinary labels (white sticky) will serve if they are fixed in place with polytape. Small parcel labels, the tie-on type, are useful for large packages.

HOW TO PACK

The most important points in packing are to eliminate as much air as possible from the package and to seal it thoroughly. A sign that too much air has been allowed to remain, or get, in is heavy frost inside the package.

Pack in quantities that will be used up for one meal. It is much more convenient to thaw several small packages than one large one. The maximum-sized container for quick thawing and heating is 1 quart (1 l).

HOW TO USE SHEET MATERIALS

Use a generous sized piece of material and put the food in the centre of the sheet. Wrap as tightly as possible in exactly the same way as a parcel. All folds and ends should be sealed with polytape except when foil has been used for wrapping. This can be sealed by pressing firmly.

An alternative method for wrapping odd-shaped parcels is the butcher's method. Place the food diagonally in one corner of the paper, fold the corner over the food, fold in the sides, roll up and seal.

USING BAGS

Open the bag completely to avoid air pockets forming in the folds. Pack solid food in as tightly as possible, pressing out air pockets. The bags should then either be heat-sealed or the top gathered together in a bunch and tied securely with plastic-coated wire or a pipe cleaner. To do this effectively put the wire round once and then fold over the bunched top of the bag and secure by twisting the wire round the double bunch. If there is plenty of room at the top of the bag it can be tied in a knot instead of fastened with wire.

USING BOXES AND TUBS

If the food is liquid or semi-liquid, a space must be left between the top of the food and the lid to allow for the expansion that will take place during freezing. The amount of space to leave depends on the food. None is needed for baked goods or for foods that have air spaces between the pieces, e.g. loose vegetables, bony meat. For dry packs or semi-soft foods leave $\frac{1}{2}$ inch (1 cm); watery food 1 inch ($2\frac{1}{2}$ cm), or $1\frac{1}{2}$ inches ($3\frac{3}{4}$ cm) in glass jars.

To eliminate air pockets, shake down loose foods; press down and smooth moist ones. If liquids are added after the food has been put in the container, swirl several times to ensure penetration of the liquid.

Unless they are guaranteed airtight, waxed carton lids should be sealed with polytape.

OVERWRAPS

These are used for protecting the original wrapper from damage when food is moved about in the freezer. A roll of mutton cloth is the best. Simply cut off a generous length, insert the package and knot the mutton cloth at each end. The label can be put between package and overwrap. Inexpensive string or nylon bags or clean nylon stockings can be used and if they are of different colours are a help in identifying foods.

LABELLING

All packets need to be labelled because even with a transparent wrapper it is often difficult to identify the contents when frozen. Other important information should also be included, e.g. raw or cooked; with or without sugar. The name should be printed in large clear letters for quick reading.

If it is not possible to write on the package itself, use a sticky label, either a special freezer one or one held on with polytape, or a tie-on label.

PENCILS

The best are "felt" pencils, or wax pencils (Chinagraph), or wax crayon. A ball-point pen is suitable for either labels covered with polytape, on the tape itself, or on non-waxed wrapping paper.

To use different-coloured pens or labels is a help in identifying packages in the freezer.

It is useful to keep a note-book or card index to record details of your own experiments for future reference, as well as for any details of additions to be made to the food when it is used, methods of thawing, etc.

Chapter Six

CLEANING AND CARE OF THE FREEZER

THE FORMATION OF FROST

Frost is due to moisture being deposited on the coldest part of the freezer, usually where the freezing pipes and coils are situated. A certain amount of this moisture comes from the packets themselves, the amount depending on how efficient the packing is. Most of it comes from warm, moist air entering the freezer every time the door is opened.

Cold air cannot hold as much moisture as warm air. When warm air enters the freezer and is cooled a stage is reached at which it is saturated with moisture. Further cooling results in the moisture being deposited as frost.

The rate of formation of frost depends on the number of times the door is opened and on the amount of moisture in the air. This can be considerable in a steamy kitchen or in warm, wet weather.

Less warm air enters with a chest-type freezer than with the upright type and the formation of frost is usually less.

DEFROSTING

The best time to do this is when stocks are running low. For example, if you freeze your own garden produce, defrost before the summer harvest begins. Otherwise do it in the winter because then it is easier to keep the remaining stock cold while out of the freezer for the defrosting process.

The freezer will usually need to be defrosted once or twice a year, though the front-opening type may need more frequent defrosting if the door is often opened.

In the past much advice has been given on this subject and the freezer owner was often told the freezer must be defrosted as soon as the frost was an eighth of an inch thick. The reason for this was the belief that the formation of frost interfered with the efficient running of the freezer. This is thought today to be less important.

It is wise, however, to scrape off the frost when it becomes about $\frac{1}{4}$ inch ($\frac{1}{2}$ cm) thick. This will delay the formation of thick ice and the need for defrosting.

The scraping should be done with a wooden or plastic spatula, taking care that nothing sharp which might damage the freezing coils is used. The frost should be collected and removed from the freezer. With some types of freezer it may be necessary to remove some of the food during this process, but the operation should take only a few minutes.

Defrosting should be carried out when the ice gets so thick that it is a nuisance: for example, when it takes up valuable freezer space or prevents the easy removal of drawers or trays. This does not usually happen until the ice is from $\frac{1}{2}$–1 inch thick (1–2$\frac{1}{2}$ cm).

If the instruction book provided with your freezer gives instructions for defrosting, follow that. Otherwise proceed as follows.

HOW TO DEFROST

1. Switch off the freezer.

2. Remove all food. If there is room in the refrigerator stack the food there with the control at its coldest point. Otherwise, wrap the frozen food in several layers of newspaper, pack it in a pile in a cool place and cover it with a clean blanket.

3. Scrape out any frost using a blunt wooden or plastic tool.

4. Leave the freezer door open and place bowls of hot water inside to hasten defrosting. Leave until the ice begins to loosen. If the freezer has no drain for the water, line the bottom with layers of newspaper to catch the ice and water.

5. Scrape again to remove the ice.

6. Wash the walls of the cabinet with hot water in which 1 Tbs bicarbonate of soda to 1 quart (1 l) of water has been dissolved. Rinse well and dry gently with a clean tea towel. Leave the door open until the cabinet is quite dry, otherwise ice will very quickly form again.

7. Should there be any stale smell after normal cleaning of the inside, wash the freezer a second time with a solution of $\frac{1}{2}$ pint ($\frac{1}{4}$ l) vinegar to 1 gallon warm water (4$\frac{1}{2}$ l).

8. If there are removable racks wash and dry them and return them to the dry freezer.

9. Switch on again and let the freezer run for not less than 30

minutes before returning the food. Dry the outside of any container damp with condensation.

10. The outside of the freezer is kept clean by wiping with a damp cloth, if necessary with a little detergent on it. Wipe dry with a tea towel and occasionally polish with a silicone dressing.

SELF-DEFROSTING FREEZERS

Some modern freezers are self-defrosting, sometimes once every 24 hours. An electric timer automatically starts and stops the defrosting daily. This can be a disadvantage if wide temperature fluctuations result (see page 16).

Chapter Seven

USING THE FREEZER

1. Before using a new freezer, wash it out with warm water containing baking soda in the proportion of 1–3 tablespoons to 1 quart of water (1½ l). Rinse with clean water and dry with a clean cloth. Close the door, switch on and leave for 4 hours before loading with frozen food, or 12–24 hours before doing any freezing.

2. Freeze food at below −18°C (0°F) if possible and store not higher than this. For fast freezing place packages next to the freezer lining, or over the freezer coils on the special shelves in an upright freezer.

3. The maker usually says how much food can safely be frozen at a time. This is given as either one-tenth of the total cubic capacity, or as 2–3 pounds per cubic foot capacity in 24 hours (1–1½ kg per 28 litres).

It is a mistake to try to freeze more than this as the freezing process will then be slowed down, but, more serious, the temperature of the freezer will be raised and cause reduction in the quality of the foods already frozen. It is a help if the foods to be frozen are first chilled in the refrigerator, or, in the case of vegetables, are chilled in ice-water after blanching and before packing.

4. Avoid placing packets of food to be frozen in contact with

those already frozen. This, too, will cause deterioration in quality by raising the temperature of the frozen food.

If the freezer has a separate compartment for freezing food, clear this of already frozen food before using it. If the freezer is an upright model with freezer shelves it is a good plan to reserve one shelf for this purpose.

As far as possible, avoid stacking packages to be frozen as this slows down the process. If you can allow about 1 inch (2½ cm) between packages this is a help to speedy freezing.

6. Use good packing materials (see page 24) and pack food in amounts to consume at one meal. It is useful to pack in units of 2, 3, or 4 portions, depending on the size of the family. When more portions are wanted, use several packs. This is better than freezing large amounts of any one food, giving a better-quality product and being more practical for catering.

It is always good policy to handle only small amounts of food at a time, the quantity you can get ready and frozen quickly. This applies to both raw and cooked foods.

7. When you freeze a food for the first time, make a note of how you did it and what the results were like, in case you want to make an alteration to the method next time. If you use the food at intervals make a note of the quality of the last lot, that is, how long it keeps in good condition.

8. Do not waste freezer space by storing food which can equally well be kept in the refrigerator, for example, butter and other fats, unless you want to carry a supply to last for months. Neither is it sensible to use the freezer to store foods which are more satisfactorily preserved by some other method, for example, tomatoes and pears.

WHAT TO FREEZE

There are very few foods that cannot be frozen satisfactorily, but which are best for you to freeze depends on what you want the freezer to do (see page 19).

Start with the foods that seem likely to be the most useful to you and make a critical appraisal after a few months. Are you just putting foods in because the freezer is there or are all the foods stored a real help with catering? Does the way you are using it really save you time and make catering easier?

Remember that food does not improve during frozen storage

and it is therefore pointless to freeze food unless it serves a useful purpose.

In the following pages you will find the pros and cons of freezing various foods discussed under the appropriate headings.

BULK BUYING FROZEN FOODS

Frozen foods can be purchased in bulk from a number of sources such as large stores and specialist firms. This is a great convenience and can lead to considerable savings provided you bulk-buy foods you use frequently. Most firms offer a wide range of cooked and raw foods. What constitutes an economical buy varies with individual needs.

If you store bulk-purchased foods for a long time the savings you have made can be offset by the cost of storage unless that particular food happens to rise sharply in price after you have purchased your stock. A rapid turnover is the ideal to aim at.

With bulk buying the method of delivery is very important. If the food is allowed to thaw in transit then the saving in money will be at the expense of quality. My experience has been that firms specialising in frozen food deliveries are on the whole better than the large store. There must, of course, be someone at home to receive the goods and put them straight into the freezer.

TRANSPORTING FROZEN FOOD

If you are transporting frozen food for any distance it pays to buy or make an insulated hold-all. It is also possible to buy Freezella/Thermella sachets which are stored in the freezer. On removal they retain their cold for a long time and can be put in the shopping hold-all. Put the sachet on top of the packages— cold falls. It is also a help to wrap the frozen food in newspaper, especially if you are not using an insulated container.

If you are not able to keep the food well frozen during transit use it up fairly soon as fluctuating temperatures reduce quality.

HOW TO FREEZE PRE-COOKED FOODS

A very large number of pre-cooked foods can be frozen successfully. By cooking two or three times the amount needed for one meal and freezing the rest, considerable preparation time can be saved. Or you may like to devote an evening or part of a day to cooking dishes for a week or more ahead. Favourite recipes using seasonal foods are also a good investment, for example, fruit pies and tarts, fruit desserts, mint sauce for winter lamb.

PREPARATION OF THE FOOD

1. Be very careful to practise good kitchen hygiene during the preparation and cooling processes.
2. Avoid over-cooking, rather cook until barely tender, remembering that the food will cook some more during the thawing and heating.
3. As some flavours become more pronounced or change character with freezing, it is a good plan to add only half the required amount of salt, pepper and spices, and the remainder during thawing, whenever practicable.
4. Chill the cooked food quickly by standing the cooking vessel in one containing ice-cold water. Finally, chill in the refrigerator before freezing.
5. To preserve flavour and quality, good packing is essential. Pack the food solid to exclude as much air as possible. Covering the food with sauce or gravy also helps to exclude air.
6. If the food is frozen in a wide, shallow container, the thawing and heating will take less time than if the same amount of food is put into a narrow, deep container.
7. Label the package carefully.

FREEZING AND STORING

For rapid freezing, chill the dish in the refrigerator before freezing it. The shorter the storage time of pre-cooked foods, the better the quality, so aim to use it up within 1–2 months, preferably less.

HEATING AND SERVING

Foods to be served cold should be thawed in the refrigerator whenever possible. But for speed, thaw at room temperature and then refrigerate until ready for service.
Foods to be served hot should be heated as rapidly as possible as this preserves the quality.

Thawing and heating can be done in a moderately hot oven E 400°F G 6. This will take about 1 hour for a quart-sized casserole. It is difficult to give precise times because so much depends on the nature of the food and the kind and shape of the container. The advantage of freezing food in small packs is obvious. Do not use a lid or the heating time will be longer. When the food begins to thaw use a fork to separate pieces gently to speed thawing. Sauces should be stirred vigorously to make them smooth.

For small amounts of food, a double boiler is very satisfactory and quicker than the oven. Stir to break up food and smooth sauces.

If you have a cooker where hot-plate control is good at low temperatures, and if you possess a wide, heavy pan, thawing and heating foods (in a sauce) over a direct gentle heat is the quickest way of all. The food needs to be watched closely until liquid begins to run, then break up gently, finally stirring to make the sauce smooth. The heat can be gradually increased to speed things up.

In the sections dealing with individual foods and recipes I have indicated the suitable methods of thawing.

ROTATION OF STOCK

Unless you are storing farm or garden produce for long periods, use the freezer as a place to keep foods for 2–3 weeks or a month or two. That is, aim at having a quick turnover of stock. In this way you get the maximum use from the freezer and the foods are of higher quality than if stored for a longer time.

Hang up a list of the contents of the freezer, and strike off those taken out. In this way you can see when things need replenishing without having to look through the frozen stock. Some careful housewives recommend noting the dates when items are put in and taken out, but this is not necessary when you plan to have a quick turnover. Nor do I think it necessary for garden produce unless the harvest times of a specific food are very spread out.

A further point in favour of a quick turnover is that it reduces the cost of each food, which is the initial cost plus the cost of storing.

Following is an example of the way I keep my freezer record on a pad hanging beside it.

38

CHICKEN (3 lb) 3, 2, 1 (meaning there are 2 left)
LIVER PÂTÉ (4 oz size) 5, 4, 3, 2, 1, (meaning there is only 1 left)

The list is grouped into kinds of foods:
Meat, Poultry, Fish, Vegetables, Fruit, Pastry, Bread and so on.

THE STORAGE LIFE OF FOODS

Any advice on this matter should be taken only as a general guide because so much depends on the quality of the food before freezing and on the many other factors already discussed.

Recommended storage times are usually based on a storage temperature of −18°C (0°F) which is not allowed to fluctuate more than 4°C (5°F) above or below this. Under these conditions the product should undergo no detectable deterioration in quality. This assumes it is also properly thawed and cooked.

Exceeding these storage times does not mean the food will become unsafe to eat, but it may lose flavour and there may also be texture changes which are undesirable to the palate.

When purchasing commercially frozen food, the safe storage time may be much less than these given here, due to wrong handling from factory to consumer.

As already pointed out, a quick turnover of stock is the wise policy for all frozen foods.

STORAGE LIFE AT −18°C (0°F)

Food	Time
MEATS (RAW)	
Bacon	1 month
Offal	3 months
Ham	3–4 months
Game, pork, veal, lamb	6–8 months
Venison	8–10 months
Poultry	10–12 months
Beef	12 months
FISH (RAW)	
Shellfish	1 month
Fish	3–6 months
VEGETABLES (RAW)	8–12 months
FRUIT (RAW)	6–9 months

Food	Time
DAIRY	
Cheese	4–6 months
Ice-cream	1 month
BAKED GOODS	1–2 months
PRE-COOKED MEALS	1–2 months

WHAT TO DO IF THE FREEZER STOPS WORKING

If the freezer is in the kitchen you will probably miss the sound of the motor working at regular intervals and realise something is wrong. If the freezer is in the cellar or garage you will only know when you go for supplies.

Before calling the service engineer check the following points:

1. Make sure the mains supply to the house is still on.
2. Make sure the freezer is still plugged in and that the switch has not been turned off in error.
3. Check to see that a fuse has not blown.

DO NOT OPEN THE FREEZER DOOR UNTIL AN HOUR OR TWO AFTER THE SUPPLY HAS BEEN RESTORED.

The length of time the food will remain frozen depends on how much of it there is in the freezer—the more, the longer. A well-filled freezer maintains −18°C (0°F) for 36–48 hours and a higher, though still safe, temperature for about 24 hours longer. If the freezer is only partly filled the food may stay frozen for less than 24 hours. A chest type remains cold longer than an upright model.

A 4–6 cubic foot (112–168 litre) freezer with a full load is not likely to begin to spoil in less than 3–4 days, a larger one in 5 days, always provided the door is not opened.

It is possible to insure against loss of food in freezers and, if you keep a large stock of expensive food, you may think it worth while doing this.

RE-FREEZING FOOD

This is not advisable except as a salvage operation, and when a large amount of food is involved. Even then it should not be done if the food is completely thawed, because spoilage is likely to have started. But if there are ice crystals, or if the centre of the food is still frozen, it should be safe to re-freeze. Use the food up as soon as possible after this.

Re-freezing of partially thawed food is not itself harmful but the quality of the food will be affected, probably in colour, flavour and texture.

WHAT TO DO WITH COMPLETELY THAWED FOOD

Vegetables It is best to discard them because they will be so poor in colour and flavour as to be unpalatable. If there are still some ice crystals present in the vegetables, cook and use for soups or purées, both of which could be re-frozen.

Meat, Fish and Poultry If the food is still cold (as from a refrigerator) it should be safe to cook it. Then use it at once or re-freeze. Thawed, cooked meat, poultry or fish should always be discarded as most unsafe to use.

Fruits You will be able to tell by the look of them if they are usable. They can often be salvaged by cooking, then use or re-freeze.

Bread, Cakes, Pastry (not pies), Biscuits These can all be safely re-frozen.

Chapter Eight

STOCKS, SOUPS AND HORS D'OEUVRE

The freezer is very useful for storing complicated soups which take a long time to prepare. Make enough for several meals and store the surplus. As soups take up a lot of freezer space make them with the minimum possible liquid (the pressure cooker is ideal for this), and then dilute when re-heating. Ingredients like potatoes, rice, pasta and seasonings are best added when thawing and heating the soup.

It is also worth while to freeze good-quality stock or consommé. Boil it first to condense it, cool quickly, and either put it in a container or in ice trays to make cubes. Then transfer the frozen cubes to bags.

For thickened soups use cornflour, rice flour, rolled oats or fine oatmeal. These give better results than flour, especially for long storage periods.

41

Chilled soups are a good choice for the freezer. Thaw them in the refrigerator and serve while still cold.

To Thaw and Heat Hot Soups
Thick soups should be thawed and heated in a double boiler. Stir or beat thoroughly to make smooth. Clear soups can be thawed over a direct gentle heat. Cubes of frozen stock or consommé can be used to make individual portions of soup by adding water or milk; or use them frozen in place of a meat cube or ordinary stock. Remember that you have previously condensed the stock, and water should be added when it is used.

GARNISHES FOR SOUP
It is very useful to have a supply of these at hand in the freezer.

Croûtons
These can be fried or toasted and frozen in small bags. Alternatively freeze cubes of bread to fry as required. They can be fried while still frozen. The latter is the better method for long storage periods as fried foods tend to become rancid with time.

Julienne Vegetables
Vegetables cut in matchstick-thin slices can be frozen in small bags for garnishing. For short-term storage there is no need to blanch them first.

Herbs for Garnishing
See page 128.

Grated Parmesan Cheese
This normally keeps well in the refrigerator but if you want to grate enough to last a long time, store some in the freezer in small bags or boxes. It thaws very quickly at room temperature but give it plenty of time, to restore full flavour.

CHICKEN OR DUCK STOCK
You will only want to make this if you think that home-made stock is better than a chicken cube or if you do not want to waste a good chicken carcase.

It is usually better to bone cooked poultry before freezing so there is often a carcase to dispose of. If you have a pressure cooker stock can be made from the carcase quite quickly.

COOKING TIME: $\frac{1}{2}$ hr pressure cooking: QUANTITIES for $1\frac{1}{2}$–2 pt (1 l)

1 *chicken or duck carcase:* 2 *pt water (4 c or* $1\frac{1}{4}$ *l):* 1 *stick celery:* 1 *onion:* 1 *carrot:* 1 *bay leaf.*

Wash and cut up the celery; scrape and slice the carrot; skin and slice the onion; wash the bay leaf. Put all ingredients in the pressure cooker and cook for 20–30 mins. Alternatively simmer in a pan for 2 hrs. Strain. The stock can be further concentrated by boiling rapidly.

To Freeze
Cool as quickly as possible by standing the pan in ice-cold water. Pour the stock into rigid containers or into an ice tray to make cubes. When the cubes are frozen pack them into a polythene bag.

To Use
Use in place of white stock in any recipe. Thaw and heat over a direct gentle heat or add frozen to hot liquid.

Some Suggestions for Using the Stock
1. Make a sauce by thickening the stock with some roux from the freezer (see page 50). Season to taste with salt and pepper.
2. Use the sauce with some frozen vegetable purée to make a quick soup, or use a purée of canned vegetables. Thin with cream or evaporated milk.
3. Make a quick curry soup by using 2 eggs and 2 teaspoons curry powder to 2 pints of stock (1 l). Heat the stock to boiling. Beat the eggs and curry together. Remove the stock from the heat and beat in the egg mixture. Serve at once.
4. Make a soup by thickening the stock with a little roux. Season to taste and add a few sliced mushrooms and something green like a few shredded Brussels sprouts, some chopped parsley or watercress.

BORTSCH

COOKING TIME: 1 hr: QUANTITIES for 8–12

12 oz onion (360 g): 1 lb raw beetroot (½ kg)

Skin the onion. Wash, trim and peel the beetroot. Shred both of them on a coarse grater.

2 Tbs sugar: 2 Tbs vinegar: 2 pt brown stock (4 c or 1¼ l)

Put in the pan with the vegetables and boil gently for about 20 minutes.

12 oz cabbage (360 g)

Wash and shred the cabbage. Add it to the soup and boil for a further 20 minutes.

4 Tbs tomato paste: salt and pepper to taste

Add to the soup together with some more stock if needed. Season sparingly. Cook until the cabbage is tender.

To Freeze
Stand the pan in ice water to cool quickly. Ladle the soup into rigid containers, seal and freeze.

To Use
Add 2 pt (1¼ l) more stock for the full recipe. Thaw and heat to boiling over a direct heat. Break up with a fork as it begins to thaw.

To Serve
Sour cream, yoghurt or evaporated milk with lemon juice. Put a spoonful in each plate as the soup is served.
Optional extra—sliced frankfurter sausage.

CREAM SOUPS, VEGETABLE
These can be made completely and stored, or quickly made from stored frozen ingredients which are:

Vegetable Purée
This can be frozen purée, see page 122, fresh vegetable purée, canned vegetables drained and sieved or put in the blender.

Thin Béchamel Sauce
Fresh made or frozen (see page 50).

Stock
White stock (see page 42) or chicken cubes.

Cream or Evaporated Milk and possibly butter as well.

Garnish
Chopped herbs from frozen stock (see page 128), or slices or dice of the same vegetable as forms the basis for the soup.

Proportions

QUANTITIES for 8

1½ pt béchamel sauce (3 c or ¾ l): ½ pt vegetable purée (1 c or ¼ l): ½ pt white stock or milk (1 c or ¼ l): Thin cream: seasoning and garnish.

Add the hot purée to the sauce and whisk to make smooth. Add stock or milk and cream, re-heat, season to taste, garnish and serve hot or freeze.

To Freeze
Stand the pan in ice water to cool quickly, stirring occasionally. Pour into freezer boxes, seal and freeze.

To Use
Heat in a double boiler, beating well to make smooth.

CRÈME DE VOLAILLE PRINCESS (Cream of chicken soup)
A soup to make and store for special occasions.

QUANTITIES for 8

1½ pt thin Béchamel sauce (¾ l) (see page 50) or *Velouté sauce* (see page 60): ½ pt chicken purée (1 c or ¼ l): ½ pt chicken stock or milk (1 c or ¼ l): salt and pepper: 3 Tbs cream

To make the chicken purée put cooked chicken in an electric blender with enough stock to make a smooth purée. Otherwise rub the chicken through a sieve.

Add the purée to the hot sauce and add stock or milk to give the desired consistency. Taste for seasoning. Add the cream.

To Freeze
Stand the pan in cold water to chill, stirring the soup occasionally. Pour into rigid containers leaving a good head space, seal and freeze.

To Use
Thaw and heat in a double boiler. Serve garnished with white asparagus tips and chervil.

Variations
The same recipe can be used to make other soups, substituting a vegetable purée for the chicken but using the same sauce and stock.
Crème de Celeri using fresh cooked or canned celery
Crème d'Asperges Vertes using canned green asparagus
Crème d'Artichauts using canned artichoke bottoms or hearts.

GASPACHO (Spanish Cold Soup)

QUANTITIES for 8

1½ *pt canned tomato juice (3 c or ¾ l): 1 lb cucumber (½ kg): pinch pepper: pinch dried garlic: 2 Tbs olive oil: 4 Tbs wine vinegar: 2 Tbs red wine: sugar and salt to taste: chopped parsley*

Peel the cucumber and grate it coarsely. Mix all the ingredients together, using parsley to garnish.

To Freeze
Pour into rigid containers leaving a good head space. Seal and freeze.

To Use
Thaw in the refrigerator or at room temperature, taste for seasoning, and serve while still ice cold.

46

OXTAIL SOUP

COOKING TIME: 4½ hrs or 1 hr pressure cooking: QUANTITIES for 4–6 for the soup; the meat will give 3–4 portions.

1 *ox tail*

Ask the butcher to cut it into joints. Remove any surplus fat, wash the joints and dry them.

2 *carrots*: 1 *small turnip*: 1 *large onion*: 2 *stalks celery*

Peel or scrape the carrots, skin and slice the onion, peel and chop the turnip, wash and slice the celery.

1 *oz fat* (2 *Tbs or* 30 *g*)

Heat this in a large saucepan and fry the oxtail in it until brown all over. Alternatively fry it in the pressure cooker. Remove the oxtail from the pan and fry the vegetables until they begin to brown. Return the meat to the pan.

1 *bay leaf: pinch ground mace or nutmeg:* 1 *sprig parsley: salt and pepper:* 6 *peppercorns:* 4 *pt hot water or* 2 *pt for the pressure cooker* (2½ *or* 1¼ *l*).

Add these to the pan, bring to the boil and simmer or pressure-cook for the required time. Strain. Cut up some of the bits of meat from the small joints and add these to the soup. Stand the pan in ice water to cool rapidly. When cold, remove fat from the top.

To Freeze
Pour the soup into freezer boxes, seal and freeze.

To Use
Turn into a saucepan and thaw over a direct gentle heat. Bring to the boil, taste for seasoning and add a small glass of sherry or red wine.

To Use the rest of the oxtail
Remove the meat from the bone and use it to make a mince or shepherd's pie by adding a good brown sauce (see page 54).

47

HORS D'OEUVRE

The preparation of these can be time-consuming but certain types of hors d'oeuvre may safely be prepared several days in advance and frozen. Freeze each kind separately in bags or boxes. Varieties which are delicate should be frozen on trays before packing in boxes with Cellophane paper between the layers.

To Use
Thaw in the refrigerator or at room temperature and then store in the refrigerator until required.

SUITABLE HORS D'OEUVRE FOR FREEZING
Sliced sausage such as salami and liver sausage.
Smoked salmon, smoked eel, buckling, sliced raw kipper.
Sliced cold meats, diced or cut in strips and dressed with oil and vinegar.
Shrimps and prawns and small tubs of potted shrimps.
Cold cooked vegetables with french dressing, for example, cauliflower, beetroot, celery, onion, french beans, leeks and raw sliced mushrooms.
Sardine butter, see page 62.
Chicken Liver Pâté, see page 113.
Liver Terrine, see page 102.
Potted Duck, see page 118.
Kipper Pâté, see page 82.
Marinated Herrings, see page 82.
Ratatouille, see page 133.
Grapefruit, see page 142.
Fruit juices, see page 137.

Chapter Nine

SAUCES

Most sauces can be frozen satisfactorily. Mayonnaise and its derivatives are exceptions. They tend to separate and are not satisfactory for freezing except when very small amounts are used with other foods.

It is worth while making large amounts of complicated and time-consuming sauces. Freeze the surplus in small containers, the most satisfactory size being about $\frac{1}{2}$ pint (1 c or $\frac{1}{4}$ l). It is advisable to season lightly and add more during thawing.

Use leak-proof containers, chill the sauce quickly and freeze at once.

I also find it very useful to make more than enough for one meal of fairly simple sauces and store the surplus. This means that one can use sauces more freely than time might otherwise permit and this helps to improve the quality of daily menus.

To Use

Reheat frozen sauces in a double boiler or over a very gentle direct heat. Stir and break up as soon as some liquid appears at the bottom of the pan. Stir vigorously or whisk frequently to make smooth. It may be necessary to thin the sauce before serving.

Egg yolks and cream can be added at this stage but I find it more practical to add them before freezing. Seasoning will need adjustment.

Adjusting Recipes and Methods

Sauces which are added to food before freezing sometimes have a tendency to separate or curdle during thawing and heating. This is because it is not always possible to stir the sauce during heating without spoiling the appearance of the food.

When making sauces for this purpose any of the following adjustments may be made:

1. Make sure the roux of fat and flour or other thickening has been thoroughly cooked before adding the liquid.

2. Thicken with cornflour instead of wheat flour, using only half the amount of cornflour as you would for flour.

3. Use rice flour to replace half or all of the wheat flour.

4. Make the sauce with only three-quarters of the required amount of flour. Mix the remainder to a paste and stir it into the finished sauce before adding the sauce to other foods. The additional raw thickening will cook as the dish is thawed and heated.

5. Use fine oatmeal in place of wheat flour for savoury dishes.

6. Be careful not to add too much fat. Use the minimum needed to make the roux. Fat increases the tendency to curdle.

ROUX

It is a great convenience to have a stock of roux for thickening sauces instead of having to make it fresh every time.

8 *oz butter* (240 *g*): 9 *oz cornflour* (270 *g*) or use *potato flour or arrowroot*.

Melt the butter in a small, thick pan, add the cornflour or alternative, thickening gradually, and mix well. Heat and stir until it becomes semi-liquid.

For a white roux cook for a few minutes, stirring all the time. For a brown roux cook very gently, stirring frequently, until the roux is a pale brown colour. This cooking may be done in a moderate oven.

Leave the roux to become cool and set but not hard.

To Freeze

Put small heaps of the roux (about 1 oz or 30 g) on pieces of foil. When they are quite cold wrap each piece in the foil and put them all in a bag. Seal and freeze.

To Use

It can be used frozen or thawed. Drop the roux into the liquid to be thickened, and stir vigorously until it thickens. A small whisk used for stirring helps to keep the sauce smooth.

This is a very good way of thickening liquid in a casserole or similar dish.

The roux will keep in good condition for many months.

BÉCHAMEL SAUCE

COOKING TIME: 15–20 mins: QUANTITIES for 4 × ½ pt (¼ l) containers.

1 *shallot or small onion: 1 small carrot: 1 piece of celery: 1 bay leaf: 5 peppercorns: 2 pt milk* (1¼ *l*)

Peel the onion or shallot and clean the other vegetables. Put them in a pan with the milk and bring to the boil. Remove from the heat, cover and leave to infuse for 5 minutes. Strain.

2 *oz butter or margarine* (4 *Tbs or* 60 *g*): 2 *oz cornflour* (6 *Tbs or* 60 *g*): *salt*

50

Melt the butter or margarine and stir in the cornflour. Cook gently for 3–5 minutes, stirring all the time. Remove from the heat and gradually add the strained milk, whisking until the sauce is smooth. Return to the heat and stir until it boils. Simmer gently for about 5 minutes, stirring occasionally. Season lightly.

¼ pt single cream (½ c or 1½ dl)

Stir in the cream.

To Freeze
Put the pan in a container of ice water to chill quickly. Stir occasionally to prevent a skin from forming. When the sauce is cold spoon it into the containers, leaving ½ inch (1 cm) head space. Seal and freeze.

To Use
Put the frozen sauce in a double boiler and heat until thawed. Beat well to make smooth. Make sure it is properly heated and taste for seasoning. If necessary, thin with a little milk or cream or some stock.

Ways of Using Béchamel Sauce
When you have a stock of sauce in the freezer it is possible quickly to make any of the following sauces. Quantities for 1 pt (2 c or ½ l).

Anchovy Sauce
Add 1 Tbs anchovy essence.
Optional extras, some chopped capers or a little lemon juice.

Egg Sauce
Add 2 chopped hard-boiled eggs.

Caper Sauce
Add 2 Tbs chopped capers and a little of the vinegar.

Mornay or Cheese Sauce
Add 4 oz grated cheddar cheese (1 c or 120 g) or use half cheddar and half parmesan.
Add 1 tsp made mustard.

Mussel or Oyster Sauce
Add 2 doz cooked or canned mussels or oysters, 1 tsp anchovy essence and 1 tsp lemon juice.

Mustard Sauce
Add 2 tsp mustard blended with 1 Tbs vinegar.

Parsley Sauce
Add 2 Tbs chopped fresh or frozen parsley.

Shrimp Sauce
Add 4 oz chopped or canned shrimps and 4 oz whole shrimps.

Soubise or Onion Sauce (for ½ pt Béchamel)
Stew 8 oz (240 g) chopped onions in 2 oz (60 g) butter or margarine until tender. Add the sauce and 1 tsp sugar. Heat in a double boiler. Rub through a sieve or put in the blender. Re-heat, taste for seasoning and serve hot.

BOLOGNESE MEAT SAUCE

COOKING TIME: 1½ hrs: QUANTITIES for 6–8

2 oz cooked ham (60 g): ½ oz butter (1 Tbs or 15 g): 2 oz bacon (60 g)

Remove the rind from the bacon and chop bacon and ham into small dice. Alternatively mince coarsely.

Heat the butter and fry the meat until it begins to brown.

1 small onion: 1 small carrot

Skin the onion, scrape the carrot, and chop both into small dice. Add to the meat and continue cooking until the vegetables are slightly softened.

4 oz boneless veal (120 g): 12 oz lean beef steak (360 g): 4 oz boneless pork (120 g)

Chop the meat finely or mince it coarsely. Add to the pan and cook, stirring frequently, until the meat no longer looks red.

¼ pt stock (½ c or 1½ dl): ¼ pt white wine (½ c or 1½ dl)

Add to the meat and simmer until the liquid has almost evaporated.

2 *Tbs tomato paste: salt and pepper*

Add tomato and seasoning to the meat with enough hot water almost to cover it. Cover the pan and simmer slowly, preferably in the oven, until the mixture is thick.

To Freeze

Chill the mixture quickly by standing the pan in ice water. Put in rigid containers, allowing head space. Cover and freeze.

To Use

Thaw and heat in a double boiler or over a very gentle heat, stirring frequently. Before serving, bring to the boil to make sure it is hot.

Optional Additions before serving are:

A pinch each of ground cloves and nutmeg and $\frac{1}{4}$ pt cream ($\frac{1}{2}$ c or $1\frac{1}{2}$ dl).

Serve as a sauce with spaghetti or noodles or use to make Lasagne Bolognese.

LASAGNE VERDI BOLOGNESE

COOKING TIME: 25 mins plus time to thaw frozen sauces: QUANTITIES for 4

3 *oz green lasagne* (90 *g*): $\frac{1}{2}$ *pt Bolognese sauce* (1 *c or* $\frac{1}{4}$ *l*): $\frac{1}{2}$ *pt Béchamel sauce* (1 *c or* $\frac{1}{4}$ *l*): 1 *oz grated parmesan cheese* ($\frac{1}{2}$ *c or 30 g*)

Cook the lasagne according to the direction on the packet. Thaw and heat the sauces separately.

When all are ready grease a shallow baking dish and put a layer of meat sauce on the bottom, then a layer of Béchamel sauce, a sprinkling of cheese and a single layer of cooked lasagne.

Repeat these layers, ending with sauce and cheese.

Bake in a moderate oven for about 25 minutes.

ALTERNATIVE METHOD

The dish may be assembled as already described and then chilled and frozen.

To Use

Heat, uncovered, in a moderate oven for ¾–1 hour or until bubbling hot and the cheese browned on top.

BROWN SAUCE

This is useful to have in the freezer for a number of purposes including the making of réchauffé dishes such as Shepherd's Pie and Mince or to make a meat sauce for spaghetti, (add mince and tomato paste). For Mushroom Sauce see page 62.

COOKING TIME: ½ hr: QUANTITIES for 4: ½ pt (¼ l) containers

4 medium sized onions: 4 small carrots

Skin the onions, peel or scrape the carrots. Cut both into small pieces.

3 oz fat (90 g)

Heat the fat in a saucepan and fry the vegetables until they are beginning to brown.

3 oz flour (9 Tbs or 90 g)

Add the flour, stir well and cook, stirring frequently, until the flour browns, about 15–20 minutes.

2 pt brown stock (4 c or 1¼ l): sprig of parsley: ½ bay leaf: salt and pepper.

Remove the pan from the heat, stir in the stock. Return to the heat and stir until the sauce boils. Add the bay leaf and parsley. Boil gently for ½ hour. Strain and season to taste, or leave the seasoning until the sauce is heated for use.

To Freeze

Cool the sauce quickly by standing the pan in ice water. Stir occasionally as it cools. Spoon into the containers, seal and freeze.

To Use

Thaw and heat in a double boiler or in a pan over a direct gentle heat. As liquid begins to run, stir to break up the sauce, finally beating to make it smooth. Taste for seasoning, bring to the boil and serve.

Optional Additions
Wine and/or tomato paste.

CRANBERRY SAUCE

COOKING TIME: 10 mins: QUANTITIES for 8 or more

1 *lb cranberries* (4 *c or* $\frac{1}{2}$ *kg*) 1 *pt water* (2 *c or* $\frac{1}{2}$ *l*)

Pick over the cranberries, removing any bad ones and stalks. Wash and put in a pan with the water. Boil, crushing them with a spoon during cooking. When they are quite tender, rub them through a sieve or cool a little and put in the blender.

8 *oz sugar* (1 *c or* 240 *g*)

Add the sugar to the purée, reheat and stir until it dissolves. Strain to make sure it is quite smooth.

To Freeze
Put the sauce in small rigid containers. Cool as quickly as possible and then seal and freeze.

To Use
To serve hot, heat in a double boiler or over direct gentle heat, thinning with water as necessary.

To serve cold, thaw at room temperature or in the refrigerator. For serving with meat, turn out of the container like a mould. For using as a sauce, e.g. over ice-cream, break up the sauce, thinning with water as necessary.

CURRY SAUCE

COOKING TIME 35–40 mins: QUANTITIES for 4: pt ($\frac{1}{4}$ l)

2 *oz fat* (60 *g*): 2 *medium onions*

Skin and chop the onions and fry them brown in the fat.

2 *oz cornflour* (6 *Tbs or* 60 *g*): 4 *Tbs curry powder*

Add to the onion and mix and cook for 2–3 minutes.

2 *pt stock* (4 *c or* 1$\frac{1}{4}$ *l*)

Add the stock gradually and stir until it boils.

4 *small apples: Grated rind and juice of* ½ *lemon:* 2 *tomatoes:* 2 *tsp brown sugar:* 2 *tsp salt:* 1 *bay leaf:* 2 *Tbs chutney*

Peel and chop the apples, chop the tomatoes. Add to the sauce with the other ingredients and simmer for 30 minutes. Rub through a sieve.

To Freeze

Stand the container in iced water to chill quickly, stirring occasionally. Pour the sauce into rigid containers, leaving a head space. Seal and freeze.

To Use

Turn into a double boiler or a pan over a very low heat. Break up as it thaws and beat to make smooth. Make sure it is properly hot before serving.

CURRIED EGGS

QUANTITIES for 4

½ *pt curry sauce* (1 *c or* ¼ *l):* 4 *hard boiled eggs:* 8 *oz rice* (1 *c or* 240 *g)*

Boil the eggs, shell and cut in half. Thaw and heat the sauce. Heat the eggs in it. Boil the rice.

Serve the eggs and sauce in a border of rice.

CURRIED PRAWNS

QUANTITIES for 4

1 *pt shelled fresh or frozen prawns* (2 *c or* ½ *l):* ½ *pt curry sauce* (1 *c or* ¼ *l):* 8 *oz rice* (1 *c or* 240 *g)*

Thaw and heat the sauce, add the prawns and make hot. Boil the rice.

Serve the prawns and sauce in a border of rice.

CHICKEN CURRY See page 112.

CURRIED COD'S ROE

QUANTITIES for 4

½ pt curry sauce (1 c or ¼ l): 4 portions cod's roe (12 oz or 360 g): 8 oz rice (1 c or 240 g): lemon wedges

Thaw and heat the sauce. Heat the roe in the sauce.

For cooking fresh cod's roe, see page 80. Canned roe is suitable for this recipe.

Boil the rice.

Arrange the roe in a border of rice and pour the sauce over it. Garnish with lemon wedges.

ESPAGNOLE SAUCE

COOKING TIME: 2 hrs: QUANTITIES for 6 or more

2 oz butter (6 Tbs or 60 g): 2 oz chopped ham or bacon (60 g)

Heat the butter in a saucepan and fry the ham or bacon in it for a few minutes.

1 medium onion: 1 small carrot: 6 mushrooms or equivalent in stalks

Peel the onion and scrape the carrot. Chop them. Wash and chop the mushrooms. Add the vegetables to the pan and continue frying until they begin to brown.

1 oz cornflour (3 Tbs or 30 g)

Sprinkle the cornflour on to the vegetables, stir and cook until it begins to brown.

1 pt brown stock (2 c or ½ l); or water and meat cubes

Add to the pan and stir until it boils. Boil gently for an hour.

2 Tbs tomato paste

Add to the sauce and continue cooking for another hour, adding more water if it seems to be getting too thick. Strain the sauce through a fine sieve.

⅛ pt sherry or madeira (¼ c or ¾ dl): Salt and pepper

Add the wine and season to taste.

To Freeze
Put the pan in a bowl of ice-cold water to cool the sauce quickly.
Pour it into rigid containers, seal and freeze.

To Use
Heat in a double boiler or over a direct gentle heat. Beat well during heating to make the sauce smooth.

This sauce may be used for covering meat to be frozen and then re-heated with the meat, for example cooked sliced tongue or sweetbreads, or use it as a brown sauce in any recipe.

MINT SAUCE
Dry mint leaves can be frozen, then chopped frozen to make mint sauce in the usual way. If you want to freeze the mint sauce ready made, the following ways give good results.

METHOD 1
   ½ *pt chopped mint* (1 *c or* ¼ *l*): 2 *oz caster sugar* (4 *Tbs or* 60 *g*): *vinegar*

Use enough vinegar just to moisten the leaves, stir to dissolve the sugar. Freeze in small containers, leaving a good head space. Dilute with more vinegar to thaw and serve. The colour may be lost slightly but the flavour is very good.

METHOD 2
As above without any vinegar which is added for thawing. The mint keeps a better colour but the flavour perhaps not quite as good.

METHOD 3
As Method 1 but mix the chopped mint with boiling water just to moisten. Cool and freeze. Add vinegar to thaw. Probably the best method for both colour and flavour.

MUSHROOM SAUCE
   1 *pt brown sauce* (½ *l*) see page 54. OR *Béchamel sauce*, page 50. OR *Velouté sauce*, page 60.
   1 *doz button mushrooms, fresh or frozen*

Wash and chop the mushrooms and simmer them in the sauce for 5 minutes. Taste for seasoning and serve hot.

To Freeze
Chill quickly by standing the pan in ice water. Spoon into containers or use to mask cooked meat, poultry or fish to be frozen.

To Use
Thaw and heat in a double boiler. For the sauce frozen with meat, poultry or fish, heat the dish, uncovered, at E 400°F G6 for ¾–1 hour.

MUSTARD SAUCE

COOKING TIME: 10 mins: QUANTITIES for 8

4 oz butter or margarine (120 g): 2 oz flour (6 Tbs or 60 g)

Melt the fat in a saucepan and mix in the flour. Cook for a minute without allowing the flour to colour. Remove from the heat.

2 Tbs dry mustard

Add to the pan and mix until well combined.

4 Tbs tarragon vinegar: ½ pt water or fish stock (1 c or ¼ l)

Stir into the flour mixture until smooth. Return to the heat, stir until it boils and simmer for 5 mins.

To Freeze
Stand the pan in iced water to cool quickly, stirring frequently to prevent a skin from forming. Pour into rigid containers.
    Seal and freeze.

To Use
Heat in a double boiler or over a very gentle direct heat. Beat to make smooth.

SAUCE SUPRÊME

QUANTITIES for 4–6

1 pt Velouté sauce (2 c or ½ l): see page 60. Chicken stock (use
59

cube and water if none available): $\frac{1}{4}$ pt single cream ($\frac{1}{2}$ c or 1$\frac{1}{2}$ dl): 3 egg yolks: Seasoning

Thaw and heat the velouté sauce in a double boiler. Add chicken stock to make it a thin sauce.

Mix the egg yolks and cream. Add a little of the hot sauce to them, mix well, return to the pan and stir until the sauce thickens. Taste for seasoning and serve.

TOMATO SAUCE

COOKING TIME: 35 mins: QUANTITIES for 2 pts (1$\frac{1}{4}$ l)

3 oz fat (90 g): 2 onions: A few pieces of bacon or bacon rinds

Skin and chop the onions. Heat the fat and fry the onion and bacon rinds in it until the onion begins to brown.

2 oz flour (6 Tbs or 60 g): 1 pt stock (2 c or $\frac{1}{2}$ l): 1 pt tomato juice (2 c or $\frac{1}{2}$ l)

Add the flour to the pan and mix well, cooking for a few minutes. Remove from the heat and gradually stir in the liquid. Return to the heat and stir until it boils.

1 bay leaf: 1 tsp sugar

Add to the sauce and boil gently for 30 mins. Strain.

To Freeze
Stand the pan in ice water to cool quickly. Pour the sauce into $\frac{1}{2}$pt boxes, seal and freeze.

To Use
Thaw and heat in a double boiler or over a direct gentle heat. Season to taste with salt and pepper. Serve hot.

VELOUTÉ SAUCE
To make and store when there is a chicken carcase to use up.

COOKING TIME: $\frac{1}{2}$ hr pressure cooking for the stock, see page 41. 10 mins for the sauce: QUANTITIES for 2 pt sauce (1$\frac{1}{4}$ l)

2 oz butter or margarine (4 Tbs or 60 g): 2 oz cornflour (6 Tbs or 60 g): 2 pt chicken stock (4 c or 1¼ l): ground nutmeg or mace: salt and pepper

Melt the fat and stir in the cornflour. Stir and cook over a gentle heat until the mixture becomes smooth and runny. Cook for a minute or two longer.

Add the cold stock and stir until the sauce boils. Simmer for 5 minutes. Season sparingly.

To Freeze
Chill by standing the pan in ice water and stirring the sauce occasionally. Pour into ½ pt containers leaving a head space, seal and freeze.

To Use
Thaw in a double boiler and beat to make smooth. Use to serve with poultry or vegetables; for mushroom sauce, see page 58, or Sauce Suprême, see page 59, to serve with boiled poultry.

SAVOURY BUTTERS
These are very useful to keep in the freezer for many purposes: to serve with grilled meat or fish; to dress boiled vegetables; for baked potatoes; for sandwiches, with or without an added filling; for sandwich cakes and loaves, see page 194; for savouries and canapés; for open sandwiches; for spreading on toast.

To Freeze
The butter can be shaped in small pats or balls and packed in boxes with Cellophane paper between the layers. Alternatively shape the butter into a thin sausage about ¾ inch diameter, rolling it between two pieces of foil.

Wrap it in double foil and freeze.

To Use
Pats are put straight on the hot food, without thawing. A roll can be sliced to make pats or thawed and used for spreading or piping.

MAÎTRE D'HÔTEL OR PARSLEY BUTTER
4 oz softened butter (120 g): 2 Tbs chopped parsley: Lemon juice to taste

Beat all together to a smooth cream.

TARRAGON BUTTER

4 oz softened butter (120 g): ¼ pt tarragon leaves (½ c or 1½ dl)

Blanch the tarragon leaves by pouring boiling water over them. Then plunge them in cold water, drain and dry by pressing between paper towels. Chop the leaves, work them into the butter and finally rub through a sieve to make it smooth. This makes a well-flavoured green butter.

MINT BUTTER

4 oz softened butter (120 g): ½ pt mint leaves (1 c or ¼ l): ½ pt parsley sprigs (1 c or ¼ l)

Wash the mint and parsley leaves and boil them in the smallest possible amount of water until they are pulpy. Rub through a sieve. Work into the butter until smooth.

MUSTARD BUTTER

4 oz softened butter (120 g): 2 Tbs made French mustard

Combine thoroughly

CURRY BUTTER

4 oz softened butter (120 g): 1 tsp curry powder: ½ tsp salt: Few drops onion juice

Combine thoroughly.

ANCHOVY BUTTER

4 oz softened butter (120 g): 2 tsp anchovy essence

Combine thoroughly.

SARDINE BUTTER

Take equal weights of canned sardines and softened butter. Mash the sardines, including bone and skin, and work into the butter. Rub through a sieve to make smooth.

Add lemon juice to taste.

WATERCRESS BUTTER

Make as Maître d'Hôtel Butter, substituting watercress for the parsley.

PAPRIKA BUTTER

1 *Tbs chopped onion:* ½ *oz butter* (15 *g*): 4 *oz softened butter* (120 *g*): *good pinch paprika*

Fry the chopped onion in the ½ oz butter adding the paprika during frying. Cool. Work into the butter and then rub through a sieve. Add more paprika if liked but avoid over-seasoning.

SWEET SAUCES

APPLE SAUCE

COOKING TIME: ½ hr: QUANTITIES for 8

2 *lb cooking apples* (1 *kg*): 8 *oz sugar* (1 *c or* 240 *g*)

The usual method is to peel, core and slice the apples before cooking them, but I prefer to wash them and cut them in pieces without peeling or coring. This produces a sauce with a greenish tinge which is attractive. Whichever way you do it, cook the fruit with just enough water to prevent burning. When the fruit is quite soft rub it through a sieve to give a smooth sauce. Add the sugar, stir to melt, and stand the pan in cold water to cool as quickly as possible. Put the sauce in ½ pt (¼ l) size containers, seal and freeze.

To Use

For a cold sauce, thaw in the unopened container either in the refrigerator or at room temperature.

For a hot sauce, empty it into a saucepan and either put over a gentle heat or in the top of a double boiler or in the oven.

CHOCOLATE SAUCE

QUANTITIES for 6–8

2 *Tbs cornflour:* 3 *Tbs sugar: pinch salt:* 2 *Tbs cocoa powder:* 1 *Tbs soluble coffee:* 1 *pt milk* (2 *c or* ½ *l*)

Put the dry ingredients in a small basin and mix to a smooth paste with some of the cold milk. Heat the rest of the milk and pour into the blended mixture. Stir and return to the heat. Stir until it boils and simmer for 5 minutes.

Vanilla essence: Cream

Add flavouring to taste and thin as desired, with cream.

To Freeze
Cool quickly by standing the pan in ice water. Stir occasionally. Pour into small boxes, seal and freeze.

To Use
Heat in a double boiler or thaw in the refrigerator or at room temperature and serve cold. Thin further with more cream if necessary.

CREAM CHANTILLY
This freezes more satisfactorily than plain cream.

QUANTITIES for 10

$\frac{1}{2}$ *pt whipping cream (1 c or $\frac{1}{4}$ l)*

If the cream is to be piped use double cream, otherwise use a mixture of equal quantities of double and single cream. Whip the cream until thick but not buttery.

1 *oz icing sugar ($\frac{1}{4}$ c or 30 g): vanilla essence or other flavouring*

Fold in the sugar and flavouring.

To Freeze
Either freeze in a rigid container or pipe rosettes or put small blobs of the cream on foil on a tray. Freeze these small pieces and then pack them in a box in layers with Cellophane paper between.

To Use
Thaw the container of cream in the refrigerator and serve as a bowl of whipped cream or use in place of sweetened whipped cream in making cold sweets. The rosettes and blobs can be put

frozen on cold sweets as a garnish and will be thawed by the time the dish is served, about 15 mins.

RAW FRUIT SAUCES OR SYRUPS (To make in the blender)

Raw ripe fruit: caster or icing sugar to taste

Any fruit can be used but the juicy ones are the best, for example cherries (especially morellos), black and red currants, raspberries and strawberries.

Wash the fruit and remove stones and stalks.

Put the fruit in the electric blender with sugar to taste and blend just enough to make the juice but not produce froth. Strain if necessary.

To Freeze
Put in small leak-proof containers or freeze in an ice-cube tray, remove the frozen cubes and put them in a bag.

To Use
Thaw in the refrigerator or at room temperature. Use as a sauce for ices or cold sweets, to make drinks and in place of fruit juice in any recipe.

The frozen cubes can be put into cold water to make an iced drink.

ALTERNATIVE METHOD for a fruit sauce.
Use frozen fruit purée. Thaw and, if unsweetened, add icing sugar to taste. If necessary to dilute it add syrup made with water and sugar or honey or use the syrup from canned fruit. Liqueurs can be added to taste.

HARD SAUCE
Useful to store for steamed fruit puddings, especially Christmas pudding. It can also be used for a cake filling.

4 oz butter (120 g): 8 oz icing sugar (240 g) or 8 oz caster sugar:
1 tsp vanilla essence or 2 Tbs rum, sherry or brandy

Warm the butter slightly but do not melt it. Cream the ingredients together thoroughly and flavour to taste. Freeze in a small bowl or boxes or shape into a thin sausage about ¾ inch diameter, using two pieces of foil for shaping. Wrap in double foil.

To Use
Thaw the bowl of sauce at room temperature and serve in the bowl. Cut off pieces from the roll and put on top of portions of pudding while the sauce is still frozen.

CUMBERLAND RUM BUTTER

Make as above but use fine light brown sugar instead of the icing sugar, add ground nutmeg and cinnamon to taste and flavour with rum.

MELBA SAUCE

COOKING TIME: 10–15 mins: QUANTITIES FOR 8–12

1½ lb fresh or frozen raspberries (750 g)

Cook without any water until the fruit is reduced to a pulp. Rub through a sieve.

1 Tbs cornflour or potato starch

Blend this with a little cold water. Stir into the purée and heat until it thickens, stirring all the time. If cornflour is used, simmer for 2–3 minutes: potato starch is cooked as soon as it thickens.

Sugar: lemon juice

Sweeten and flavour to taste. Cool.

To Freeze
Put in jars or rigid containers, leaving a head space. Seal and freeze.

To Use
Thaw in the refrigerator or at room temperature. Use as a sauce for ice-cream or other cold sweets.

Chapter Ten

DAIRY PRODUCE, EGGS AND FATS

MILK

With normal milk deliveries there is not much point in using the freezer to store it but it is useful sometimes to have an emergency supply.

Freezing brings about changes in milk. The freezing, storing and thawing processes cause a breakdown of the fat emulsion and the milk protein becomes less stable. There is thus a tendency for the cream to separate in globules when the milk is thawed while the change in protein may cause flocculation. This is less likely to happen with homogenised milk provided it is not stored more than one month.

As milk has a high water content it is important to leave plenty of head space for expansion, not less than $1\frac{1}{2}$ inches ($3\frac{3}{4}$ cm). Wax cartons are the most satisfactory containers. Put the carton in a polythene bag for extra protection.

Thaw frozen milk at room temperature.

CHEESE

Freezing gives a crumbly texture to hard cheeses like Cheddar and blue cheeses. This does not affect the flavour and the cheese is suitable for cooking or for use in salads and sandwiches.

Freezing is a very good way of keeping ripe soft cheeses like Camembert or Valmeuse, when they have reached the right stage of ripeness. Freezing preserves them at this stage without affecting the texture or flavour. Before serving they must be given ample time for thawing (1–2 days in the refrigerator), and then allowed to come to room temperature. If this is not done the cheese will lack flavour.

Unsalted cottage cheese is satisfactory in the freezer and so is cream cheese.

To Freeze

It is better to freeze in packs of $\frac{1}{2}$ lb ($\frac{1}{4}$ kg) or less, otherwise thawing takes a very long time. The cheese should be wrapped in

67

moisture/vapour-proof paper. If it is already in a carton or foil wrapping it can be stored in this for a week or two, but for longer storage it should be over-wrapped. A polythene bag or double foil is suitable.

Storage Time
Cottage cheese 3–4 months; others 4–6 months.

CREAM
Frozen cream has a tendency to separate on thawing but this happens less with double cream. Separation may be very largely overcome by either using short storage periods, or whipping the cream lightly before serving, or adding a little sugar before freezing it. Frozen cream is suitable for any cooking purpose and for making ice-cream.

To Freeze
If the cream is to be stored for more than 2–3 weeks it is advisable to pasteurise it. To do this heat it in a double boiler to 82°C (180°F) and then cool it rapidly before freezing. Pour into suitable containers allowing $\frac{1}{2}$ inch (1 cm) head space, seal and freeze. Thaw in the refrigerator or at room temperature.

Whipped Cream
If it is to be kept for more than 2–3 weeks pasteurise it as above. Whip in the usual way, adding sugar to taste. It can then be frozen in rigid containers or shaped to use for garnishing. To do this put freezer paper on a tray and either pipe the cream on to the paper in rosettes or put in small mounds. Freeze uncovered, remove and pack in layers in a box with freezer paper between the layers. Seal and freeze. It can also be used for filled cakes or prepared sweets and frozen.

To Use
Thaw at room temperature or in the refrigerator. Put the rosettes or blobs on the top of sweets as a garnish while they are still frozen and they will thaw very quickly (about 15 mins). Pasteurised cream will keep up to 4 months.

CREAM CHANTILLY See page 64.

BUTTER

Unsalted butter made with pasteurised fresh cream will keep almost indefinitely. Any butter will keep up to 6 months. If it is already foil wrapped it can be stored for 1–2 weeks like this, but for longer keeping over-wrap with freezer paper.

Butter balls and butter pats can be frozen too and are useful items to store for special occasions and for serving on grills and fish. Seasoned butters like Maître d'Hôtel Butter can be made in quantity and frozen in pats, in an ice-cream tray or in a roll for slicing. For recipes see page 65. Hard Sauce and Cumberland Rum Butter see pages 65-6.

OTHER FATS

Lard, margarine, cooking fats and dripping keep well provided they are suitably wrapped.

EGGS

Those who keep chickens and those who are able to purchase eggs at reduced prices may want to use the freezer for storing them. Others will do it for convenience, especially for keeping whites or yolks left over during cooking.

Whole eggs freeze very well provided they are removed from the shells, which do not allow enough room for expansion during freezing and so break.

Hard-boiled eggs are not satisfactory as they tend to become very rubbery.

To Freeze Whole Eggs

Put some waxed sweet or cup-cake cases on a tray and break an egg straight into each. Cover each with a circle of Cellophane and freeze. Then pack them closely in a carton or bag. They will keep for 8–10 weeks. It is better to thaw before frying or poaching but they can be cooked frozen when they will take longer than fresh eggs.

Alternatively, beat the eggs to mix white and yolk before freezing. Small cartons are the best to use for this and be sure to label with the number of eggs.

There is sometimes a tendency for the egg to become pasty. Adding a little salt or sugar helps to prevent this. When using the eggs it is useful to know that 3 tablespoons of the thawed mixture equal one egg.

Freezing Egg Whites

Put into small containers, label and freeze. Thaw before using. They will beat up very satisfactorily, even better than before freezing. Two tablespoons thawed equal one white.

It is useful to freeze together the number you know you will need for a favourite recipe. Otherwise they are more useful if frozen singly in the same way as whole eggs, see above. For recipes using egg whites see pages 151, 152, 154, 157, 164.

Freezing Egg Yolks

Mix with $\frac{1}{2}$ teaspoon salt or 1 teaspoon sugar for each egg yolk. This prevents them from going pasty during storage. Put in small containers and label clearly whether sweet or salted. Thaw before using. For recipes using egg yolks see pages 59, 133, 153, 158, 187, 191–2. One tablespoon equals 1 egg yolk.

General Notes on Thawing

Thaw in the unopened container in the refrigerator or at room temperature. If they have been frozen in a watertight container they can be thawed quickly by standing the container in warm water. After thawing, egg whites will keep for several days in the refrigerator.

Chapter Eleven

FISH

As every fishmonger, fisherman and housewife knows, fish is very susceptible to spoilage. This is brought about by bacterial contamination and by enzymes within the cells of fish muscle. To reduce spoilage by bacteria, clean handling of the fish at all stages is important while, to delay enzyme spoilage, it is necessary to freeze fish as soon as possible after catching and to store at low temperatures. Frozen fish is affected more by poor storage conditions than any of the other frozen foods.

There is an additional problem with oily fish. Fat tends to

become rancid due to oxidation during storage and thus the quality of fatty fish is not retained for as long as that of white or lean fish.

The problem the British fishing industry has to face is that of having to go long distances to the fishing grounds and the difficulty of getting the fish quickly to the freezing plants. One solution, but a costly one, is to have freezing factories at sea acting as mother ships to trawlers. The best quality frozen fish is processed in this way, where it is caught.

Frozen fish is an important convenience food in modern catering. Provided the consumer is particular about its purchase and storage and does not try to keep stocks for too long, it is a very high quality food.

If the packaging has been poor or has become damaged freezer "burn" due to moisture evaporation can result. This shows as a matt white appearance of the fish which becomes very light in weight. Milder forms of deterioration cause a stringy dry texture of the flesh.

HOW TO FREEZE FRESH FISH

If the amateur fisherman wants to freeze his own surplus catch he should do it as soon after landing as possible. In addition he should store it at the lowest possible temperature, preferably below 18°C (0°F).

The fish should be scaled, gutted and prepared as it would be for cooking. It can be frozen whole, filleted or in steaks. Individual pieces of fish should be frozen with a double layer of Cellophane paper between each piece for easy separation while it is still frozen.

If it is to be kept for any length of time the quality is often better if the fish is glazed with a thin layer of ice. This excludes air and helps to reduce evaporation.

To Glaze

Freeze the fish (whole or large pieces) on trays. Then dip the fish quickly in ice cold water and re-freeze. Repeat this operation until there is a layer of ice about an eighth of an inch thick.

Freezing in a Block

Small fish and fillets can be frozen in an ice block. Place the fish in a rigid container, cover with cold water and freeze. If the container is wanted for other purposes, dip it in warm water to loosen the block, remove and wrap.

Large Whole Fish
These can be stuffed with freezer paper to keep them in a good shape, e.g. salmon.

Shellfish
It is even more important that these should be very freshly caught. Shellfish can be frozen raw, in or out of the shell, but it is more practical to freeze the cooked fish, taking care not to overcook.

Freeze cooked crab and lobster meat packed tight in containers; or the meat can be packed in the shell and frozen for short periods.

Cooked prawns and shrimps can be shelled and packed tight in containers.

If the shellfish is going to be eaten without further cooking clean handling during shelling and packing is very important.

Freezing Fishmonger Fish
Really fresh fish purchased a few days in advance of requirements can be frozen, as already described, but it will not keep its quality for long. Wrap closely to exclude air. I find fillets of firm, close-textured fish are especially satisfactory, for example rock salmon or huss.

BUYING FROZEN FISH
Reject any packages which are broken or misshapen. This indicates poor storage conditions. There should not be loose snow or ice in the package, nor should there be any discoloration of the flesh. Patches of matt white on the surface indicate freezer "burn" and poor quality.

Mis-shapen packages indicate that the fish has been allowed to thaw and then re-frozen, another indication of poor quality.

When thawed, frozen fish is judged for quality in the same way as fresh fish: good odour, good colour, firm flesh.

THAWING
Commercial packers of frozen fish usually recommend cooking it while still frozen. If fillets and steaks are completely thawed they

tend to lose a great deal of moisture as "drip", and with this, flavour and some nutrients. Fish only needs to be thawed if it is to be coated with batter for deep frying or if the pieces have to be separated. Partial thawing is usually sufficient.

Ideally thawing should be done in the refrigerator, allowing 6 hours per pound, more if the fish is thick. Shellfish should be thawed slowly in the refrigerator. Then use straight away.

All fish deteriorates faster after it has been frozen than when it is fresh.

Fish can be thawed more quickly at room temperature or by putting the packet in cold water but this quicker thawing gives a poorer flavour and a greater loss of "drip".

COOKING FROZEN FISH

In general, methods are the same as for fresh fish, except that it is advisable to carry out the cooking more slowly at a lower temperature. Cooking time may be a quarter as much again or even twice as long depending on the thickness of the fish and the temperature used. If the flakes separate easily when tested with a fork, the fish is done. It is best to test a piece in the centre to make sure it is cooked right through.

To Serve

Keep a good supply of frozen savoury butters and sauces to serve with the fish and some frozen mushrooms, tomatoes and other vegetables suitable for cooking with the fish or for garnishing.

FRIED FROZEN FISH

METHOD 1 Partially or completely thaw the fish. Dip it into flour before coating with egg and crumbs or with batter. Then fry in the usual way in shallow or deep fat.

METHOD 2 This is the better way with completely frozen fish. Coat it in seasoned flour and fry gently in a little butter or oil or half and half. Fry until golden brown. See also Fish Meunière, page 76, and Plaice or Sole with Orange, page 76.

GRILLING FROZEN FISH

Brush the fish with oil or melted butter and season with salt and pepper. Place in the bottom of the grill pan, cook gently until lightly browned and the flesh flakes when tested with a fork in the middle of the piece.

TIME: 10–15 mins depending on the thickness.

Keep a supply of savoury butters to serve with grilled fish, see pages 61-3.

ALTERNATIVE

Use the following basting sauce instead of the butter or oil.

QUANTITIES for 4

2 *Tbs oil: 1–2 tsp lemon juice: ¼ tsp salt: pinch of pepper*

Mix together and brush over the fish during cooking. For variety add to it any one of the following:

1 *tsp french mustard: 1 tsp Worcester sauce: 1 tsp anchovy sauce:*
1–2 *Tbs wine: 1 Tbs ketchup*

BAKED FROZEN FISH

This is a very good way of cooking it, especially when there is an accompanying sauce to keep it moist: or when it is cooked en papillote, see recipes below and page 77.

STEWED FROZEN FISH

These dishes are usually more successful with frozen fish than with fresh because there is less tendency for it to disintegrate, see recipes below and page 77.

COOKING FROZEN SHELLFISH (raw)

Put the frozen fish straight into boiling salted water and cook for the usual times; prawns 2–3 mins; lobster ½–¾ hr depending on size; crab, 15 mins per lb.

BAKED FISH WITH WINE

This is an excellent way of cooking any of the frozen fish.
COOKING TIME: 20–30 mins: TEMPERATURE: E 375°F G 5.

Individual portions of frozen fish: Vegetables such as chopped onion, sliced mushroom, chopped tomato: Salt and pepper

Put a layer of prepared vegetables in the bottom of a baking dish large enough to allow the fish to be in a single layer. Season the fish with salt and pepper.

White wine or cider: Butter or margarine

Add sufficient liquid to moisten the vegetables and put a few knobs of butter or margarine on the top. Bake, basting once or twice during cooking. Test one piece of fish in the middle to see if it is cooked through. Lift out the fish and keep hot.

Roux

Thicken the liquid with a little frozen roux, see page 50. Serve the fish coated with the sauce.

CURRIED FISH

COOKING TIME: 20–30 mins: QUANTITIES for 4

1½ *lb frozen cod or haddock fillets (750 g): 1 large onion: 1 Tbs curry powder: 1 oz butter or oil (2 Tbs or 30 g): salt*

Cut the frozen fish in 2 inch pieces. Skin and chop the onion. Fry it in the butter or oil until lightly browned, adding the curry during the frying.

Add the fish and sprinkle lightly with salt.

Cover and cook slowly for 15–20 mins, shaking the pan occasionally to prevent sticking.

Chopped parsley: lemon wedges: boiled rice

Serve sprinkled with chopped parsley and garnished with lemon. Serve with rice.

ALTERNATIVE

Use cold as an open sandwich; or cold with salad; or as part of an hors d'oeuvre.

FISH EN PAPILLOTE

COOKING TIME: 30 mins: TEMPERATURE E 400°F G 6

This is one of the best ways of cooking frozen salmon steaks and other choice fish such as turbot. It is also very good with cod, hake and haddock.

Put each portion of fish on a piece of foil large enough to wrap it completely.

Season the fish with salt and pepper and a squeeze of lemon juice.

For fish needing additional flavour, add a little chopped mushroom or onion and/or a few herbs.

Make a loose parcel with the foil. Stand the parcels in a baking dish and bake.

Open one to test if the fish is done.

Serve with any liquid in the parcel or add the liquid to an accompanying sauce.

FRIED FROZEN FISH À LA MEUNIÈRE

COOKING TIME: 5–15 mins depending on thickness

Frozen fillets, steaks or small whole fish: milk: flour: salt and pepper: butter

Coat the fish with seasoned flour. Heat the butter and fry the fish gently until it is lightly browned and cooked through. Drain and keep hot.

Add more butter to the pan, about ½ oz per person, and heat this until it is nut brown. Pour over the fish and serve at once. Garnish with parsley and lemon wedges.

FRIED FROZEN FISH WITH ORANGE

This is especially good with the small whole plaice, but can also be done with fillets.

COOKING TIME: 8–10 mins: QUANTITIES for 4

4 small whole frozen plaice or equivalent in fillets: Seasoned flour: 1 oz butter (2 Tbs or 30 g)

Coat the fish with flour and fry in the butter until the fish is brown and cooked through. Drain and keep hot.

1 large orange

76

Prepare this while the fish is being cooked. Peel and remove all pith. Cut in thin rounds, removing pips. Arrange on the fish.

1 *oz butter* (2 *Tbs or* 30 *g*)

Heat until light brown, pour over the fish and serve at once.

FROZEN FISH WITH ANCHOVY SAUCE

COOKING TIME: 10–15 mins for the fish; 20 mins for the sauce.
QUANTITIES for 4

½ *pt frozen Béchamel sauce* (1 *c or* ¼ *l*), *see page* 50: *4 frozen cod or other fillets, in individual portions:* 1 *oz anchovy fillets* (30 *g or half a small can*): *lemon juice*

Thaw and heat the sauce in a double boiler. Whisk to make it smooth. Chop the anchovy fillets and add them to the sauce together with lemon juice to taste.
Grill or fry the fish.
Serve the fish with the sauce poured over it.
If liked use more anchovy fillets to garnish the dish.

HUNGARIAN FISH STEW

COOKING TIME: 15 mins: QUANTITIES for 4

1½ *lb frozen cod or haddock fillets* (750 *g*): 1 *medium onion:* 1 *oz butter or oil* (2 *Tbs or* 30 *g*)

Cut the frozen fish in 2 inch pieces. Skin and slice the onion. Heat the butter or oil and fry the onion in it until almost tender.

2–3 *Tbs tomato paste:* ½ *tsp paprika pepper*

Add to the onions and cook for a minute or two longer. Add the fish, cover, and cook until the fish is cooked through. Shake the pan occasionally to prevent sticking.

PORTUGUESE BAKED FISH

COOKING TIME: 30 mins: QUANTITIES for 4
TEMPERATURE: E 400°F G 6

1–1½ *lb frozen fillets* (500–750 *g*): 1 *small onion:* 1 *tsp salt: pinch pepper:* 4–6 *tomatoes:* 1 *oz butter or margarine* (2 *Tbs or* 30 *g*)

Grease a shallow baking dish large enough to allow the fish to lie in a single layer. Arrange the fish in it and sprinkle with salt and pepper.

Skin and chop the onion and sprinkle it over the fish.

Skin and chop the tomatoes, or use an equivalent amount of canned ones. Add to the dish and dot with the butter or margarine.

Bake. Test the centre of one piece to make sure it is cooked.

3 oz grated cheese ($\frac{3}{4}$ c or 90 g)

Sprinkle the cheese over the fish and either put it under the grill to melt the cheese or put the dish at the top of the oven.

FREEZING COOKED FISH

Many fish recipes can be prepared in advance and frozen. This is the best way of dealing with fish which has been purchased a day or so in advance of requirements.

Cooked shellfish dishes are also satisfactory for freezing.

FREEZING FRIED FISH

Coat the fish with batter or egg and crumbs in the usual way. Fry and drain thoroughly on absorbent paper. Pack with Cellophane paper between the pieces. Wrap, seal and freeze.

To Use

Heat in a little butter in a frying-pan for 10–15 mins or in a moderate oven for 15 mins. Test a piece to make sure it is well heated.

FISH CAKES

Make these in the usual way, coating them with egg and dried crumbs. Freeze on trays and then pack with Cellophane paper between layers. Wrap and seal.

To Use

Heat a little butter in a frying pan. Fry the fish cakes over a moderate heat for 5 minutes each side or until hot in the middle.

FISH PIE

Make in a foil pie dish. Freeze. Then cover with double foil.

To Use
Heat at 400°F G 6 until brown and hot.

FILETS DE SOLE À LA FERMIÈRE

This dish is unusual in combining red wine with sole. It is an excellent type of dish for the freezer and could be made with plaice or other white fish fillets of a firm variety.

COOKING TIME: 35 mins: TEMPERATURE E 325°F G 3: QUANTITIES for 4

4 large fillets of sole or 8 small ones: ⅔ pt red wine (1⅓ c or ⅓ l): thyme, chervil, parsley.

Skin the fillets and fold them in half. Season lightly. Put them in a flat baking dish with the wine. Add a sprig each of thyme, chervil and parsley. Cover and bake gently for about 20 mins or until barely cooked. Avoid over-cooking. Lift out the fish and put it in a foil dish. Strain the wine into a small pan and boil rapidly until it is reduced by about half.

1 oz roux (30 g): 2 oz butter (60 g)

Thicken the sauce with the roux, boil for a few minutes. Remove from the heat and add the butter, stir to melt. Pour the sauce over the fish.

1 oz butter (30 g): 8 oz mushrooms (240 g)

Wash, dry and slice the mushrooms. Cook them gently in the butter until almost done. Arrange them round the fish.

To Freeze
Cool as quickly as possible, seal with a double foil lid. Freeze.

To Use
Heat in the foil dish, uncovered at E 400°F G 6 for 45 minutes.

FISH MOUSSE

Useful for cold suppers and buffet parties. It is best made with fresh or canned salmon, canned tuna, cooked smoked fish, smoked or fresh cod's roe.

1 lb fish without bone ($\frac{1}{2}$ kg): 2 Tbs tarragon vinegar: $\frac{1}{4}$ tsp paprika pepper

Mash the cooked fish or, better still, put it in an electric blender to make it smooth. Add the vinegar and paprika.

1 oz gelatine (30 g): $\frac{1}{4}$ pt hot water ($\frac{1}{2}$ c or $1\frac{1}{2}$ dl)

Dissolve the gelatine in the water and mix it with fish. Cool.

$\frac{1}{4}$ pt whipping cream ($\frac{1}{2}$ c or $1\frac{1}{2}$ dl)

Whip the cream and fold it into the fish mixture. Taste for seasoning.

To Freeze
Pour into a mould or freezer box, chill, seal and freeze.

To Use
Thaw in the refrigerator for several hours or overnight. Dip the mould in warm water to loosen and turn out on to a serving dish. Garnish with salad vegetables and lemon wedges.

FRESH COD'S ROE
Most fishmongers sell cod's roe ready cooked, and it keeps very well frozen in this state.

To cook it for yourself, tie the raw roe in a piece of muslin or similar cloth. Put it into boiling salted water and cook it very gently for $\frac{1}{2}$ hour. Lift it out of the pan, but leave it in the cloth to become cold. Cool as rapidly as possible. Cut it in half-inch-thick slices.

To Freeze
Pack the slices in a container with freezer paper between slices, seal and freeze.

To Use
Thaw in the refrigerator or at room temperature, or cook while frozen. Dust with a little seasoned flour and fry gently in hot fat until it is well browned and hot through. Serve with wedges of lemon.

Curried Cod's Roe. See page 57.

KIPPER PASTIES

COOKING TIME: 30 min: QUANTITIES for 8 pasties
TEMPERATURE E 425°F G 7

4 *medium-sized kippers, with bone or filleted*

Grill the kippers or cook in any other way preferred.
Cool and flake.

2 *oz margarine* (4 *Tbs or* 60 *g*): ½ *pt milk* (1 *c or* ¼ *l*): 2 *oz flour*
(6 *Tbs or* 60 *g*)

Melt the margarine in a small pan. Add the flour and mix well.
Stir and cook for a few minutes. Remove from the heat, stir in the
milk, return to the heat and stir until it boils.

4 *oz grated cheeze* (1 *c or* 120 *g*): 1 *Tbs lemon juice: pinch of
pepper.*

Add to the sauce, together with the flaked kipper. Leave to
become cold.

1½ *lb short pastry* (750 *g*)

Divide the pastry into 8 pieces. Form each into a round and roll it
into a circle about 6 in. in diameter. Cover one half of each round
with the kipper filling. Moisten the edges, fold over and press
edges together to seal well.

To Freeze
They may be frozen raw or cooked.
Wrap in double foil, freezer paper or put in polythene bags. Seal
and freeze.

To Use
Cooked pasties: thaw at room temperature to serve cold, or heat
at E 400°F G 6 for 20–30 mins.
    Frozen raw pasties cook at E 400°F G 6 for 45–50 mins or until
the pastry is lightly browned.

VARIATION
For party snacks make very small pasties and cook for half the
above times.

KIPPER PÂTÉ

COOKING TIME: 5–10 mins: QUANTITIES for 8 or more

1½ lb kippers (750 g): or 12 oz kipper fillets (360 g)

Put the kippers in a jug deep enough to take them with a little room to spare. Pour in boiling water to cover them. Cover the top with a lid of foil and stand the jug in a warm place for 5–10 mins. Drain the kippers and allow to become cold. Remove bones and skin.

8 oz butter or margarine (240 g): 1 tsp anchovy essence: pepper: lemon juice

Warm the butter to soften but not melt. Add it to the kippers with the anchovy and put in the electric blender until smooth. Alternatively, rub the kippers through a sieve and work in the butter and anchovy. Add pepper and lemon juice to taste. Pour into a rigid container.

To Freeze
Seal containers and freeze.

To Use
Thaw in the refrigerator and store there until required. Serve with toast as an hors d'oeuvre, or use for sandwiches.

MARINATED HERRINGS

COOKING TIME: 25 mins: QUANTITIES for 6

6 fresh herrings: 1 small onion: 1 small carrot: 1 bay leaf

Remove heads and tails, clean and scale the herrings.
　　Retain any roes and put them back inside the fish.
　　Skin and chop the onion; scrape and slice the carrot; wash the bay leaf.

½ pt white wine or dry cider (1 c or ¼ l): ¼ pt white wine vinegar (½ c or 1½ dl): ½ pt water (1 c or ¼ l): 2 tsp salt: pinch dried garlic

Use a wide shallow pan which will allow the fish to lie in a single layer (or cook them in relays).

Put all the ingredients except the herrings in the pan, bring to the boil and simmer for 15 mins.

Add the herrings and poach them gently for about 10 mins, turning once if the liquid does not cover them completely.

Remove from the heat and allow the fish to cool in the liquid.

To Freeze

Arrange carefully in one or more rigid containers. Strain the liquid over the fish, seal and freeze.

To Use

Thaw in the container at room temperature and then store in the refrigerator until required. Serve for hors d'oeuvre or with salad.

RICE AND FISH MOULDS

These make a light and delicate lunch or supper dish, or they can be used as the fish course in a formal meal.

COOKING TIME: $\frac{1}{2}$ hr: QUANTITIES for 8

1 lb ($\frac{1}{2}$ kg) raw fillets white fish (fresh or smoked): 4 oz rice ($\frac{1}{2}$ c or 120 g)

Boil the rice. Mince the fish coarsely, or chop it small.

2 eggs: $\frac{1}{2}$ tsp salt: pinch of pepper: pinch of mace: $\frac{1}{2}$ pt milk (1 c or $\frac{1}{4}$ l)

Beat the eggs lightly and add the milk and seasonings.

Mix with the fish and rice. Pour into 8 small moulds which have been well greased or oiled.

Cover with lids of foil and steam for $\frac{1}{2}$ hr.

To Freeze

Stand the moulds, uncovered, in iced water to cool as quickly as possible. Freeze uncovered. Turn out of the moulds, wrap and return to the freezer.

To Use

Return to greased moulds, cover and steam for 45 mins or until well heated. Turn out and serve with a suitable sauce such as mustard, lemon, shrimp, parsley or anchovy.

SARDINE ROLLS

QUANTITIES for 16–18 rolls: TEMPERATURE E 400°F G 6
COOKING TIME: 25–30 mins

Short pastry recipe, page 172. 16–18 sardines: Grated cheese: curry powder or lemon juice

Open the can of sardines and drain off the oil. If some of the fish are very fat they should be halved lengthwise so that all will be about the same size.

Roll the pastry into a rectangle about ¼ in thick, keeping the edges as straight as possible. Then cut in strips as wide as the sardines are long. Put a sardine on the end of a strip, sprinkle it with a little grated cheese or curry or lemon juice and roll until the pastry encloses it, cut the pastry. Unroll a trifle and moisten with water, press back in place. Repeat.

To Freeze
Place the rolls in a rigid box or on a foil tray, cover with foil or seal the box, freeze at once.

To Use
Put the frozen rolls on a baking tray, brush with milk and bake until lightly browned. Serve hot or cold for party savoury or with salad for a light meal.

TUNA LOAF

COOKING TIME: 45 mins: TEMPERATURE E 350°F G 4
QUANTITIES for 12–16

Grease two 1 lb loaf tins or line with foil or non-stick paper. The loaves can be served hot or cold or use one while hot and freeze the other.

1 lb canned tuna (½ kg): ½ pt milk (1 c or ¼ l): 6 oz fresh or frozen breadcrumbs (2 c or 180 g)

Heat the milk and pour it over the crumbs. Leave to stand.
   Mash the tuna thoroughly, including the liquid and bones.
   Add to the breadcrumbs.

4 *eggs:* ¼ *tsp pepper:* 2 *tsp salt:* 1 *Tbs lemon juice: pinch of ground mace or nutmeg.*

Separate the yolks and whites of the eggs, adding the yolks to the fish and crumbs and putting the whites in a mixing bowl. Add the flavourings to the fish mixture and stir thoroughly. Beat the egg whites until stiff and fold them into the other mixture. Put in the prepared tins and bake until firm and lightly browned.

To Serve Hot
Turn out on a hot dish and mask with a hot sauce, e.g. parsley, lemon, tomato or onion. Cut in slices and serve with vegetables.

To Freeze
Stand the tins in cold water to chill quickly. Turn out of the tin and overwrap with double foil or other refrigerator wrapping paper. Freeze.

To Use
To serve cold, thaw in the wrapper at room temperature and then store in the refrigerator. Mask with mayonnaise and serve with salad.

To serve hot, leave in the foil wrapping and heat at E 400°F G 6 for ¾–1 hr.

Chapter Twelve

MEAT

Quick frozen raw meat is available from some suppliers while most butchers will sell a side of pork or lamb suitable for freezing and cut it into joints. If you intend to store meat which the butcher buys frozen, for example New Zealand lamb, and which he normally thaws before selling, you want to ask him to sell you it still frozen hard. Re-freezing thawed frozen meat leads to loss in quality.

Quick frozen cooked meat dishes are also fairly widely available from commercial freezers.

It is simple to freeze either raw or cooked meat in a domestic freezer, and, indeed, these are among the most useful items to keep in the freezer.

CHOICE AND PREPARATION OF RAW MEAT FOR FREEZING

Very lean meat tends to dry out during freezing but the presence of an ample layer of fat and good marbling helps to prevent this. On the other hand, too much fat is undesirable as it has a tendency to become rancid. These points are very important if the frozen meat is to be stored for long periods. For short term storage, any good quality meat is suitable.

Prior to freezing, meat should be properly aged, as it would be for immediate consumption. This ensures tenderness and reduces the tendency to "drip" when the meat is thawed.

The choice of cuts to freeze depends on taste and size of family. With beef it is generally better to use the best cuts and have them boned to save freezer space. Wherever appropriate to the cooking method have joints of other types of meat boned.

Meat for Stews and Casseroles should be trimmed, cut in small pieces, and packed close in bag or container. This is more satisfactory than trying to cut frozen meat or waiting for it to thaw. Keep a supply of cut-up stewing beef, veal, and boneless pork and lamb.

Slices of Meat, Chops and Steaks should be prepared ready for cooking and frozen with Cellophane or polythene paper between pieces for easy separation while they are still frozen.

Offal is better frozen in small packs. It keeps its appearance better and thaws more quickly. Polythene bags or small rigid containers are the best to use.

Raw Minced Meat is better in small packs and in rigid containers.

Wrapping Meat not in bags or rigid containers, use the parcel wrap, see page 30, with any freezer paper. Exclude as much air as possible and under-wrap bone ends to prevent them from piercing the wrapper during storage. Seal securely and chill in the refrigerator before putting the pack in the freezer.

THAWING
Frozen meat can be cooked without thawing except when it is to be used in one of the following ways:

1. If large pieces need to be cut up small or minced.
2. If it is to be coated with flour, egg and crumbs or batter as these do not stick well to a frozen surface. Partial thawing is usually sufficient.
3. If it is to be deep fried when it is difficult to ensure that the outside is not overcooked before the inside is thawed.
4. If chops and steaks have not been separated by paper before freezing. Thaw enough to separate.
5. If the meat is to be stuffed. It is not advisable to freeze ready-stuffed meat but to freeze stuffing separately.

Whether it is better to thaw, or to cook frozen, depends on time and convenience, also on the cut. Small pieces of meat can lose a lot of moisture and nutrients in the "drip" which results from complete thawing. This is because the cut surface is large compared with the total size. In this case to cook either frozen or partially thawed is the best way.

The chief difficulty in cooking from the frozen or partially thawed state is in judging the cooking time. It is usually approximately $1\frac{1}{2}$–2 times the normal for thawed meat, but less than this for very small pieces of meat. Some guidance on this is given with cooking methods that follow.

METHOD AND TIMES FOR THAWING
The ideal is to thaw in the original wrapper and in the refrigerator. This takes considerably longer than thawing at room temperature but slow thawing gives better results because there is less "drip".

When thawed at room temperature joints of meat can appear to be thawed when the inside is still frozen. This complicates the estimation of the cooking time.

Cook meat as soon as possible after thawing as this helps to reduce the amount of "drip". Be sure to keep completely thawed meat in the refrigerator until it is to be cooked.

Approximate Thawing Times

Slices of meat, steaks and chops, approx. 1 in. thick		Joints
In the refrigerator	12–18 hrs.	3–5 hrs per lb up to 4 lb
		5–7 hrs per lb if larger
At room temp.	2–3 hrs	1–2 hrs per lb small
approx. 60°F		2–3 hrs per lb large
(In cold kitchen or larder, times can be as long as in the refrigerator)		

In an emergency joints can be thawed in cold running water or in front of an electric fan when the time for thawing will be approximately $\frac{1}{2}$–$\frac{3}{4}$ hr per lb.

COOKING

Frozen meat which has been completely thawed is cooked in exactly the same way as fresh meat and no different times or techniques are necessary.

The instructions given here are for cooking frozen meat. The recipes given are for dishes to be prepared and cooked before freezing and are some of those I have found satisfactory for short-term storage, that is, up to a month. In fact, most meat recipes can be treated in this way. Instructions which warn against freezing certain types of meat dishes are usually thinking in terms of commercial or long-term storage.

In my experience the only quality changes that take place during short-term storage are with seasoning such as pepper which can become very strong. It is always advisable to season lightly, adding more if found necessary when the dish is thawed and heated for service.

I have found some cooked cold meats tend to become dry if they are not covered with stock or a sauce, particularly low-fat meats like turkey and chicken. Mousses made with gelatine tend to "weep" and are not usually of as high a quality as those freshly made. I find the meat loaf, gelatine, terrine or pâté give better results than a mousse.

COOKING FROZEN MEAT

Roasting Frozen Meat

This only differs from roasting fresh or thawed meat in the time required. This must allow for thawing plus cooking. The time can be calculated either as 1½–2 times that usually allowed for the fresh equivalent. Alternatively allow 15–20 mins per lb extra for thawing.

The only way to be sure the cooking has been long enough is to use a meat thermometer which can be inserted half-way through the cooking time, when the meat should have thawed enough to enable this to be done. As a general rule, the larger the roast, the more additional time must be given for thawing.

Fairly slow cooking is advisable, E 300–325°F G 2–3. I find I get better results if the meat is wrapped in "Look" cooking film. This reduces the amount of juice lost by evaporation.

It is important to avoid over-cooking as this makes the joint shrink a great deal and become very dry.

Roasting Partially Thawed Meat

This can only be hit or miss unless a meat thermometer is used. As a rough guide allow 10 mins more per lb than when the meat is thawed.

APPROXIMATE ROASTING TIMES at E 325°F G 3 (Pork E 375°F G 5)

Minutes per pound

	Frozen	Thawed	Meat Thermometer
BEEF			
Unboned ribs and thin pieces	60	40	140°F rare
			160°F medium
			170°F well done
Boned and rolled and thick pieces	65	45	

(For rare meat calculate about ⅔ of these times)

| **LAMB** | 65 | 45 | 175–180°F |

Minutes per pound			
	Frozen	*Thawed*	*Meat Thermometer*
PORK			
Spare ribs and			
loin	55	35	185°F
Leg	60	40	
VEAL			
With bone	50	30	180°F
Boned and rolled	60	40	

FRYING FROZEN STEAKS

Start the cooking over a low heat to allow the meat to thaw, then, if necessary, increase the heat for browning. The only way to be sure the meat is done as you like it is to cut a slit in the steak and note the colour inside.

COOKING TIMES for steaks about ½ in thick

Times in Minutes

	Frozen	*Thawed*
Rare	9	7
Medium	11	9
Well done	13	11

For one-inch-thick steaks allow 5–6 mins longer in each case.

To Serve

Keep a stock of savoury butter in the freezer for serving with steak, see recipes page 61.

FRYING FROZEN CHOPS AND CUTLETS

Heat some oil, or half butter and half oil.

Start cooking over a low heat to allow the meat to thaw.

Use about the same amount of heat as you would to keep a pan boiling gently. If necessary, increase the heat at the end for browning. Avoid fast cooking or the meat may be very dry. To prevent fat from splashing about, rest a piece of foil across the top of the pan.

To Serve
Keep a stock of savoury butters, see page 61, or a sauce such as Espagnole, see page 57, to serve with the meat.

COOKING TIMES in minutes

	Frozen	Thawed
Lamb cutlets	12–15	7–10
Lamb chops	15–25	10–20 (depending on thickness)
Pork or veal chops	30	20
Veal Escalopes	10	4–5
Liver	10	5
Kidneys	10	5

GRILLING FROZEN MEAT

Place the meat 1–2 inches further from the heat than is normal, until the meat has thawed. Put it closer to the heat for quick browning at the end.

Cooking Times as for frying.

To Serve
Keep a stock of savoury butter in the freezer, see page 61.

BOILING AND STEWING FROZEN MEAT

It is usually more satisfactory, though not essential, to thaw the meat first. If the meat needs to be fried before stewing it is certainly better to at least partially thaw it before frying.

BRAISING AND POT ROASTING

It is usually more satisfactory to thaw first.

FROZEN PORK FILLETS IN CIDER SAUCE

COOKING TIME: 30 mins: QUANTITIES for 4
4 *portions pork fillet or boned cutlets about ¾ in thick: lard for frying*
Separate the portions of frozen meat. Heat the lard and fry the pork gently until brown on both sides. Drain off surplus fat.

91

½ pt dry cider (1 c or ¼ l): 2 Tbs tomato paste: 1 Tbs dried onion: salt and pepper

Add to the meat, seasoning lightly. Cover with a foil lid and simmer gently for the remaining cooking time, or until cooked through and tender.

ALTERNATIVE
Instead of the cider and other ingredients add ½ pt Espagnole sauce.

FREEZING COOKED MEAT

Frozen meat can be used successfully to make cooked dishes and then re-frozen.

STEWS AND CASSEROLES

These are amongst the most satisfactory cooked meat dishes for home freezing. Most recipes are suitable for short-term storage, that is, up to a month. It is advisable to use seasonings, spices and garlic lightly as these have a tendency to become more pronounced in flavour during storage. It is always possible to add more seasoning after thawing and heating.

Avoid over-cooking the meat before freezing as it will cook some more during thawing and heating. To reduce the original cooking time by about 30 minutes is all that is required.

To Freeze
All stews and casseroles should be cooled as quickly as possible by standing the cooking container in iced water and stirring occasionally.

Do not use large containers for freezing, the maximum size advisable being 1 quart. With larger containers the freezing and re-heating will take a very long time and the results are not usually as satisfactory as with smaller amounts.

The following are suitable containers:
1. Polythene bags. Stand the bag in a carton or other oblong or square container. Fill, seal and freeze. Then remove the bag from the carton for storage. This gives a more economical shape for storage.

2. Use a rigid box with a lid.

3. Freeze in the casserole, dip in water to loosen, remove and wrap. Return to the casserole for thawing and heating. Removal is easier if the casserole has been first lined with foil.

To Thaw

If a *polythene bag or rigid box* has been used, remove the contents and put in a pan over boiling water or over a very gentle direct heat. As it thaws, stir gently to break up and hasten thawing, but not to spoil the appearance.

A pint-size pack will take about 45 minutes and a quart size an hour or longer. Serve as soon as it is thoroughly heated.

If *a casserole*, return to the original container and thaw and heat in the oven, without a lid. Stir occasionally to break up and speed thawing.

TEMPERATURE: E 400°F G 6: TIME: 45 mins for 1 pt size; 1 hr or more for 1 quart size. Heat until the whole casserole begins to boil gently.

FREEZING COOKED MEAT TO SERVE COLD

The most satisfactory for this are galantines, terrines, pâtés, meat loaves and continental sausages. Sliced cold roast or boiled meats are also satisfactory for short term (up to 1 month) storage. After that they tend to become dry.

It is very important to cool and freeze cooked meat quickly and to be very particular about the hygiene of handling, packing and thawing.

It is usually more practical to slice the meat before freezing as it can then be thawed more quickly, but it keeps more moist if frozen in a piece.

It should be wrapped closely to exclude air, the parcel wrap with double foil being the best method.

To thaw, remove it from the freezer 3–4 hrs before you want to serve it and, when the slices have thawed enough to be separated, spread them out on clean absorbent paper, to remove condensation and make them look more appetising. As soon as the meat is thawed, serve it, or return it to the refrigerator until required.

Cooked joints of meat will take about the same time to thaw as raw pieces, see page 87.

FREEZING LEFT OVERS

When a large joint has been cooked, apart from freezing for serving cold, it can be made into such rechauffé dishes as croquettes or rissoles, patties and so on. These can all be very satisfactorily frozen.

CROQUETTES OR RISSOLES

Make in the usual way and freeze either before or after frying. Fried ones thaw and heat at E 400°F G 6 for 20–30 mins. Uncooked ones fry gently in shallow fat for 5–10 mins.

MEAT PIES AND PASTIES See pages 169–71.

ABERDEEN SAUSAGE (galantine for serving cold)

COOKING TIME: 2 hrs: QUANTITIES for 8–10

1 small onion: 8 oz streaky bacon (250 g): 1 lb lean beef steak or lean minced beef (½ kg)

Peel the onion. Remove rinds from bacon. If you are mincing the beef yourself, remove any fat and gristle. Mince all three together to give a fine texture. If you buy mince, chop the bacon and onion very finely before mixing with the mince.

4 oz rolled oats (1 c or 120 g): 1 Tbs Worcester sauce: 1 egg, beaten: 1 tsp salt: ¼ tsp pepper

Add these to the meat and mix well. Shape into a long, thick sausage like a breakfast sausage and wrap carefully in foil, sealing the ends well.

It can then be boiled or baked. To bake, place on a tray and cook at E 300°F G 1 or 2. To boil, put in boiling water to cover and then reduce to simmering.

Dried breadcrumbs

Remove the foil carefully and roll the cooked sausage in the crumbs. Leave in a cold place to cool as quickly as possible.

To Freeze

It may be frozen whole or sliced. If the latter, re-shape in roll form with pieces of paper between each slice. Wrap with freezer paper.

To Use

Thaw whole sausage in its wrapper in the refrigerator for several hours, or separate slices and thaw them at room temperature on pieces of kitchen paper to absorb condensation. Put in refrigerator to keep if not required for immediate service.

BEEF BOURGUIGNON

COOKING TIME: 2–2½ hrs: QUANTITIES for 8

2 lb stewing beef (1 kg): 6 oz bacon (6 rashers or 180 g): 1 medium onion: 4 oz mushrooms (120 g): a bouquet garni: ½ pt stock (1 c or ¼ l)

Cut the meat in approximately 1 inch cubes, removing excess fat. Remove rinds and cut bacon in small pieces. Skin and slice the onion. Wash and slice the mushrooms. Prepare the bouquet garni and the stock which can be meat cubes and water.

2 oz butter (4 Tbs or 60 g)

Heat the butter and fry the onion and bacon until they begin to brown. Remove from the pan, using a perforated spoon. Fry the meat until it is brown.

1 oz flour (3 Tbs or 30 g)

Sprinkle over the meat and cook until the flour begins to brown.

¼ pt Burgundy (½ c or 1½ dl): ½ tsp salt: pinch of pepper

Add these to the meat together with the mushrooms, bouquet garni, onion and bacon. Stir until it boils. Cover and cook slowly for 2 hours on top or in the oven at E 300°F G 1.

To Freeze

Stand the pan or casserole in cold water to cool the meat as quickly as possible. Turn into one or more rigid containers according to the number of portions required at a time. Leave a head space.

Alternatively, freeze in the casserole, dip in warm water to loosen the contents, remove and wrap.

To Use

Heat in the oven at E 400°F G 6, for 1–1½ hrs. For faster thawing, heat in a double boiler or in a pan directly over a gentle heat. In each case stir frequently during heating, to separate pieces of meat and allow the heat to penetrate to the centre.

BLANQUETTE DE VEAU

COOKING TIME: 2 hrs: QUANTITIES for 6

2 lb breast of veal or pie veal (1 kg): 6 small onions: 1 sprig thyme: 1 bay leaf: 2 sprigs parsley

Cut the meat in small pieces. Skin the onions. Wash the herbs and tie together with cotton.

1 oz butter (2 Tbs or 30 g): 1 oz lard (2 Tbs or 30 g): 1 oz cornflour (3 Tbs or 30 g)

Put these in a pan over a gentle heat and mix well all the time. Cook for a few minutes.

1 pt warm water (2 c or ½ l)

Stir very gradually into the pan until it boils.

Add the veal, onions, herbs and a good pinch of salt. Cover and cook very gently for 1½ hrs or until the meat is tender.

Stir from time to time. This part of the cooking may be done in the oven.

1 egg yolk: 2 Tbs cream: 2 Tbs lemon juice

Mix these in a small basin. Remove the meat from the heat. Take a spoonful of the sauce and mix it into the egg. Then pour the egg mixture into the pan, mixing well.

To Freeze

Put the pan in a bowl of cold water to cool the blanquette as quickly as possible. If you intend to store this for any length of time it may be as well to remove the onions. Either freeze in the casserole or transfer to a rigid container. Leave a head space. Seal and freeze.

To Use

Transfer to an oven-proof dish and heat at E 400°F G 6, uncovered, for 1 hour or until well heated. Check for seasoning.

BRAISED SWEETBREADS

COOKING TIME: about 45 mins: QUANTITIES for 6–8

2 lb sweetbreads (1 kg): lemon juice: salt

Put the sweetbreads in a pan with cold water to cover. Add a little salt and lemon juice. Bring to the boil and simmer for 10 mins. Drain, remove membranes and dry the sweetbreads on kitchen paper.

2 oz butter (4 Tbs or 60 g)

Heat in a pan and fry the sweetbreads until brown.

¼ pt dry white wine (½ c or 1½ dl)

Add to the sweetbreads and simmer until the wine has almost evaporated.

2 tsp cornflour: ¾ pt hot chicken stock (1½ c or 4½ dl)

Sprinkle the cornflour over the sweetbreads and add the stock gradually. Cover and simmer for 30 mins.

1 Tbs chopped parsley: salt and pepper

Add the parsley and taste for seasoning; be sparing with the pepper.

To Freeze
Stand the pan in cold water to cool as quickly as possible. Then put the sweetbreads in a rigid container, making sure there is enough sauce to cover and leaving a head space. Seal and freeze.

To Use
Heat in a double boiler or over a direct gentle heat.
   Alternatively heat in a casserole, without a lid, in the oven at E 400°F G 6 for ¾–1 hr.

CURRY

COOKING TIME 2½–3 hrs. QUANTITIES for 6–8

2 lb meat without bone (1 kg): 6 Tbs flour: 3 tsp salt.

Remove any surplus fat and cut the meat in small pieces, about ½–1 inch cubes (1–2 cm). Mix flour and salt in a paper bag and shake the meat in this until it is well coated.

2 *medium onions:* ½ *clove garlic or pinch of dried:* 2 *apples:* 1 *tomato*

Peel and chop the onions, garlic, apples and tomato.

2 *oz fat, dripping or oil* (4 *Tbs or* 60 *g*)

Heat the fat or oil and fry the meat until it begins to brown. Add the vegetables and fry until the onions begin to brown. Pour off any free fat.

2 *Tbs curry powder or more if preferred hot*

Add to the meat together with any flour left over. Mix well.

¾ *pt stock* (1½ *c or* 4½ *dl*)

Add to the pan and stir until it boils.

Grated rind and juice of ½ *lemon:* 1 *tsp brown sugar:* 1 *Tbs desiccated coconut:* 2 *Tbs raisins or sultanas*

Add to the pan together with some gravy browning if the curry is preferred a dark colour. Simmer gently on top or in the oven.

To Freeze
Cool quickly by standing the pan in iced water and stirring occasionally. Either put into rigid containers or, if it has been cooked in a casserole, freeze in the casserole. Then remove and wrap for storage.

To Use
Return to the original casserole or put in any oven-proof dish. Heat 1–1½ hrs (at E 400°F G 6) or until bubbling hot. Alternatively heat in the top of a double boiler. Stir to break up as it begins to thaw.

FRIED VEAL ESCALOPES FROZEN IN SAUCE

COOKING TIME: 3–4 mins for the veal: QUANTITIES allow 3–4 oz (90–120 g) meat and about ¼ pt (1 dl) sauce, per portion.

Fillets veal: butter and oil for frying: Espagnole sauce, page 57 or Velouté sauce, page 60 (made with veal stock), or Mushroom sauce, page 58.

Heat the butter and oil (half and half) and fry the veal until lightly browned on both sides. Drain on kitchen paper and put in a cold place to cool as quickly as possible.

The sauce should be made in advance and be cold too.

To Freeze
Arrange the fillets in a shallow box or in a casserole. Cover completely with sauce, seal and freeze. If in a casserole, remove when frozen and wrap.

To Use
Unwrap and return to the casserole or turn out of the box into a shallow baking dish. Heat, uncovered, at E 400°F G 6 for ¾–1 hr or until well heated through.

HAM OR BACON IN SPICED ORANGE SAUCE
Although both cured meat and spices can be difficult in the freezer, ham losing colour and spices becoming bitter, I find this a very satisfactory way of storing some surplus cooked ham or bacon to give an interesting meal at a later date. The sauce alone could be made and stored.

COOKING TIME: 5–10 mins: QUANTITIES for 4

12 oz cold sliced boiled or baked ham or bacon (360 g): 3 large oranges

Trim the meat to remove the fat which does not keep as well as the lean. Keep in the refrigerator while making the sauce.

Wash one of the oranges thoroughly and grate finely to give 1 Tbs of the yellow rind. Squeeze the juice from the oranges and make it up to ½ pt (1 c or 2½ dl) with water.

2 oz brown sugar (4 Tbs or 60 g): 1 Tbs cornflour: ½ tsp salt: Pinch of ground cinnamon: pinch of ground cloves

Mix these in a small pan and gradually stir in the orange juice to make a smooth mixture. Bring to the boil, stirring constantly, and cook until it thickens and clears.

1 *beef cube*

Add to the sauce and stir until melted. Simmer for 5 mins.

1 *oz butter* (2 *Tbs or* 30 *g*)

Remove from the heat and add the butter in small pieces, stirring until blended.

To Freeze
Stand the pan in iced water to cool the sauce quickly. Arrange the sliced ham or bacon in a rigid container, pour over the sauce, leaving a head space, seal and freeze.

To Use
Remove from the container and heat either over a very gentle direct heat or in a double boiler. As it begins to thaw stir very gently to separate the pieces of meat and make the sauce smooth again. Serve the meat with the sauce poured over. If necessary, the meat may be removed when it is hot, and the sauce stirred vigorously to make it smooth again.

HUNGARIAN GOULASH

COOKING TIME: 2–3 hrs or 30 mins pressure cooking: QUANTITIES for 6–8

2 *lb lean stewing beef* (1 *kg*): 2 *oz fat or oil* (60 *g*)

Trim off surplus fat and cut the meat in small pieces of ½–1 inch cubed (1–2 cm). Heat the fat or oil and fry the meat in it until it begins to brown.

2 *onions*

Peel, slice and add to the meat.

1 *Tbs paprika pepper* (*more if liked hot*)

Add to the meat and mix well

½ *pt water* (1 *c or* ¼ *l*): 1 *meat cube*: 1 *tsp salt*: ¼ *tsp carraway seeds*: 4 *Tbs tomato paste*

Add to the meat, mix well and cook until the meat is tender. Cooking may be done in a casserole in a slow oven.

To Freeze

Cool the goulash quickly by standing the pan in iced water and stirring occasionally as it cools. Transfer to freezer containers or freeze in the casserole. Remove from the casserole when the meat is frozen and wrap for storage.

To Use

Heat in the oven without a lid at E 400°F G 6 for 1 hour. Alternatively heat in a double boiler.

Stir to break up as it begins to thaw.

LAMB WITH RICE

COOKING TIME: 50–60 mins: QUANTITIES for 6–8

2 lb lean lamb without bone (1 *kg*): *3–4 Tbs oil*

Cut the meat in ½ inch cubes (1 cm). Heat the oil in a large deep frying pan or shallow saucepan. Brown the lamb in it for about 8 minutes.

2 medium onions

Skin and chop and add to the meat. Cover and cook gently without any added liquid until the onion is soft.

2 tsp salt: ½ tsp pepper: 6 Tbs tomato paste: ½ pt water (1 *c or ¼ l*)

Blend the tomato and water and add to the meat together with the seasoning. Cover and simmer for 45 mins or until the lamb is tender.

To Freeze

Stand the pan in iced water to cool as quickly as possible. Stir occasionally as it cools. Put in rigid containers leaving a head space. Seal and freeze.

To Use

Heat in the oven at E 400°F G 6 for 1–1½ hrs or until bubbling hot. Alternatively heat in a double boiler, or in a pan over a very gentle heat. Break up as it begins to thaw.

12 oz rice (360 *g*)

Cook the rice and make a circle on the serving dish with the meat in the centre.

LIVER TERRINE

COOKING TIME: 45 mins: TEMPERATURE E 400°F G 6: QUANTITIES for 6–8

8 oz liver (240 g): 1 small onion

Wash and dry the liver and remove any tubes and fibres. Skin and chop the onion.

½ oz fat (1 Tbs or 15 g)

Heat the fat and brown the liver quickly in it. Then brown the onion.

2 oz bacon (2 rashers or 60 g)

Remove the rinds. Mince liver, onion and bacon.

4 oz pork sausage meat (120g): 1 egg, beaten: 2 oz fresh or frozen breadcrumbs (⅔ c or 60 g): 1 tsp Worcester sauce: 1 Tbs lemon juice: 1 tsp celery salt: 2 Tbs red wine

Mix all the ingredients together using enough water to make a smooth soft consistency.

Line a 1 lb (½ kg) loaf tin with foil. Put in the mixture and smooth the top. Bake uncovered.

To Freeze

Stand the tin in cold water to cool as quickly as possible. Remove from the tin and wrap the terrine in double foil. Freeze.

To Use

Thaw in the refrigerator overnight or at room temperature then store in the refrigerator until required. Serve sliced with salad for a main course or as an hors d'oeuvre. It also makes an excellent sandwich filling.

MINCED BEEF LOAF

COOKING TIME: 1 hr: TEMPERATURE E 375°F G 5

QUANTITIES for 2 loaves giving 8–12 portions

    2 lb lean meat (1 kg)

This may be all beef but a better loaf is made with a mixture of beef, veal and pork. If you are mincing the meat yourself put it twice through the mincer. If the mincing is too coarse the loaf will fall apart when it is cut.

    1 medium onion: Fat for frying

Skin and chop the onion and fry it in a little fat until it is tender.

    3 oz fresh or frozen breadcrumbs (1 c or 90 g): 2 tsp salt: ¼ tsp ground mace or nutmeg: ¼ tsp pepper: 2 eggs, beaten: 2 Tbs chopped parsley

Mix all the ingredients together very thoroughly.

Divide the mixture into two and shape each loaf about 2 inches thick. The loaves may be cooked before freezing if they are required as cold meat, or they may be frozen raw. To cook, put them in a baking tin with a little fat. Cool as quickly as possible.

To Freeze

Wrap either the cooked or raw loaves in double foil, pressing it closely to keep the loaf in shape. Freeze.

To Use

Thaw the cooked loaves in a refrigerator. To speed up thawing, they may be sliced while still partly frozen. Finish thawing at room temperature, but refrigerate again if not required for immediate service.

To cook the raw frozen loaf, open up the foil to expose the top of the loaf, put foil and loaf on a tray and bake for 1½ hrs at E 325°F G 3. Serve hot with gravy or a sauce, for example tomato, Espagnole, or brown sauce, from the freezer.

PAUPIETTES DE VEAU

COOKING TIME: 1¼ hrs: TEMPERATURE E 325°F G 3

QUANTITIES for 8.

About 8 very thin escalopes of veal

Cut these into pieces approximately 4 ins by 2 ins.

    2 oz breadcrumbs, fresh or frozen (⅔ c or 60 g): 2 oz grated suet

103

(6 *Tbs or* 60 *g*): *pinch of ground mace or nutmeg:* 1 *tsp dried thyme:* ½ *tsp grated lemon rind:* 1 *Tbs chopped parsley:* ½ *tsp salt:* ¼ *tsp pepper*

Put all these in a mixing bowl.

1 *egg: milk to mix*

Beat the egg and use it with some milk to mix the stuffing to a fairly moist consistency so that it will hold together. Spread a little on each piece of veal and roll up tightly. Fasten securely with a wooden cocktail stick or with fine string tied round in two places.

1 *small onion, chopped:* 2 *oz butter or oil* (4 *Tbs or* 60 *g*)

Heat the butter or oil and fry the paupiettes until brown all over, adding the onion towards the end of cooking.

Place the meat and onion in a casserole.

1 *oz flour* (3 *Tbs or* 30 *g*): 1 *pt stock* (2 *c or* ½ *l*)

Add the flour to the fat remaining in the pan and blend well. Stir and cook for a few minutes. Remove from the heat and gradually stir in the stock. Stir until it boils.

2 *Tbs chopped parsley:* ¼ *tsp pepper:* 1 *tsp salt*

Add to the sauce, mix and pour over the meat. Cover, and cook gently until the meat is tender.

To Freeze
Cool as quickly as possible. Put in rigid containers, leaving a head space, seal and freeze. Alternatively cook in a foil-lined casserole and freeze in this. Remove and overwrap.

To Use
Transfer to a suitable dish and heat at E 400°F G 6 for 1 hr or until well heated. Remove the string or sticks and serve the paupiettes with the sauce poured over them.

While you are removing the string or sticks it is as well to cut one paupiette in half to make sure they are hot right to the centre.

PORK IN RED PEPPER SAUCE

COOKING TIME: ¾–1 hr: QUANTITIES for 8

2 lb boneless pork, leg or fillet (1 kg) cut in ½ in slices: 2 onions

If the meat is in large slices, cut it to make 8 portions. Beat well to flatten.

Skin and chop the onion.

Lard for frying: flour and salt

Dust one side of each piece of meat with flour and a very little salt. Heat enough lard for frying the meat and fry the floured side first and then the other side. As they cook, lift them out and keep them hot.

If necessary, add some more lard to the pan to fry the onions for 3 minutes.

½ tsp paprika pepper

Add during the frying of the onions.

1 oz flour (3 Tbs or 30 g)

Add to the onion and mix well.

¾ pt stock (1½ c or 4½ dl)

Stir the stock in gradually and when the sauce boils return the meat, cover and cook 20 minutes over a low heat or in the oven, until the meat is tender.

6 Tbs evaporated milk: 1 tsp lemon juice

Combine these and stir them into the sauce.

To Freeze
Stand the pan in cold water to cool the meat quickly.

Freeze in a casserole or rigid container. Leave a head space. Seal and freeze.

To Use
Either heat in a double boiler (large one needed) or in an uncovered casserole in the oven at E 400°F G 6 for an hour or more. Stir to break up as it thaws and to make the sauce smooth again.

Chapter Thirteen

POULTRY AND GAME

Quick frozen raw poultry and game have been on sale for many years. Quick frozen cooked dishes, mainly using chicken, are marketed by a number of firms.

As ready-frozen poultry is so easy to obtain there is not much point in buying birds for freezing raw at home. The possible exceptions here are game birds. Also, if you are able to get freshly killed poultry at lower prices than the quick frozen you may think it worth freezing some. I have included notes on how to do this.

Cooked poultry and game dishes are very satisfactory in the home freezer and a very useful reserve to have in stock. Practically any recipe can be used. The same general remarks apply here as I have included under the chapter on meat, see "Cooking," page 88.

The recipes included in this section are ones I have found especially useful, including some suitable for special occasions.

FREEZING RAW POULTRY

Choose young birds which are neither too fat (liable to go rancid), nor too lean (liable to go dry).

Ducklings should be about 10–12 weeks old. Young ducks of 3–7 lb weight are also suitable.

A goose should be young and tender and not more than 12 lb in weight.

Poussins, young chickens and capons are all suitable. With boiling fowl it is more sensible to freeze it as a cooked dish, or possibly in raw joints, for future casseroles.

Turkeys take up a great deal of freezer space. Plump hen birds of 8–15 lb weight are the best to freeze.

PREPARATION FOR FREEZING

Poultry should be freshly killed, not more than two days before freezing.

Game birds should be hung before freezing, to give the condition you prefer.

Hares and rabbits are best hung for 24 hours before preparing and freezing. Freezing in joints is usually the most satisfactory method, or freezing cooked dishes.

Venison should be well hung before freezing and then is treated in the same way as meat, see pages 85 to 94.

POULTRY

It is prepared as for cooking, plucked (without damaging the skin), drawn, head and feet removed, and trussed. It is better to pack the giblets separately in case you will want to cook the bird while still frozen.

Chill the birds for at least 12 hours in the refrigerator. This helps to make them more tender and speeds up freezing.

It is not advisable to stuff poultry before freezing. It can be dangerous to do this because it takes too long to freeze and thaw a stuffed bird and this allows time for the growth of food poisoning organisms. It is better to freeze the stuffing separately.

WRAPPING

Polythene bags are the most satisfactory wrapping material and special ones are made for whole birds. Before putting the bird in the bag under-wrap any sharp bones which might pierce the bag during storage and handling in the freezer. Press out as much air as possible before sealing.

For half birds, re-join them with a double layer of paper between the pieces. Then wrap as usual.

To aid separation of poultry joints when they are taken out of the freezer, pack them with Cellophane paper between the layers.

THAWING

I would only recommend cooking whole birds frozen in an emergency, and then you can only do it if they have been frozen without the giblets inside. Otherwise thaw them enough to remove the giblets.

Birds to be stuffed need to be at least partially thawed.

For frying and grilling it is better to thaw first.

Thawing is more even if it takes place in the refrigerator, or in a cold larder. Leave the bird in its wrapper. Large birds can have the thawing process speeded up by finishing them off in cold water, either running water or frequent changes.

Thawing at room temperature in the bag is satisfactory for small birds, but with large ones there may be complete thawing of the outside long before the inside is ready. This can lead to loss of a great deal of "drip" and a reduction in quality. The temperature of the room will naturally determine the speed of thawing.

If the bird is completely thawed too long before cooking there will be a considerable loss of juice and the flesh may be somewhat dry. It pays to cook poultry as soon as possible after thawing.

THAWING TIMES

	In the Refrigerator or cold larder	At room temperature (60°F)
Whole chicken	24–36 hrs.	6–8 hrs.
Chicken joints	6 hrs.	3 hrs.
Small turkey	2 days	Not recommended
Large turkey	3–4 days	Not recommended
Goose	24–36 hrs.	6–8 hrs.
Duckling and Ducks	24–36 hrs.	6–8 hrs.

Game birds similar to above depending on size.
Hare and rabbit joints as chicken joints.
Venison steaks and joints, see Meat pages 87–94.

COOKING

When frozen poultry and game have been thawed cooking methods are the same as with fresh.

ROASTING FROZEN CHICKEN

This is only possible with birds frozen ready for roasting (giblets removed), and not required stuffed.

Start the cooking at E 200°F G "Low" and allow 25 mins per lb de-frosting time; then increase the heat to E 325°F G 3 and cook for the normal time.

APPROXIMATE ROASTING TIMES

Prepared Weight	Frozen	Thawed (E 325°F G 3)
2–2½ lb	2 hrs.	1 hr.
3 lb	2¼ hrs.	1½ hrs.
4 lb	3¾ hrs.	2 hrs.

FREEZING ROAST CHICKEN TO SERVE COLD
Roast by your usual method, including stuffing if required. Drain and remove stuffing. Cool both as rapidly as possible.

To Freeze
Wrap bird and stuffing separately in double foil. Freeze.

To Use
Remove the foil and place the chicken in a polythene bag. Thaw 18–24 hrs in the refrigerator or 9–10 hrs at room temperature. Thaw the stuffing separately.

FROZEN ROAST CHICKEN TO SERVE HOT
Some people recommend re-heating a roast chicken cooked as in the previous recipe. I think the result is not as palatable as a freshly roasted chicken, having a re-heated flavour.

If you do decide to try this method heat bird and stuffing separately at E 400°F G 6 for ¾–1 hour.

FREEZING ROAST TURKEY
Cool as rapidly as possible. Remove the meat from the bones, and cut off any fat. Cooked turkey tends to become dry and is best covered with a stock or a sauce. It can then be thawed and heated in the stock or sauce, the stock being thickened with roux to make a sauce.
For serving cold freeze dry for short storage, or in stock. Thaw and drain from the stock or thaw as other cold meats, see page 93.

FROZEN FRIED CHICKEN
This takes longer to thaw and heat than to cook in the first place. The only advantage of freezing it is that last-minute frying

is avoided. Next time you cook fried chicken for dinner, cook some extra and freeze it.

Thawed frozen or fresh chicken may be used.

Remove the skin and dry the joints. Coat with egg and crumbs in the usual way. Fry in a little hot oil until the pieces are lightly browned on both sides. Drain carefully on absorbent paper and cool quickly.

To Freeze
Pack with Cellophane between the joints. Seal and freeze.

To Use
Separate the joints and place them on a baking tray.

Cook them in the oven at E 400°F G 6 for 45 minutes or until they are tender and hot.

GAME

Thaw all game in the refrigerator and then cook as fresh.

FREEZING COOKED DISHES

These are very useful for freezer storage, chicken recipes being particularly successful. I have included a few recipes for these to show the type of dish which can be frozen.

CHICKEN CASSEROLE

COOKING TIME: 1 hr for a broiler; 2½–3 hrs for a boiling fowl.

TEMPERATURE E 350°F G 4 for the broilers: E 300–325°F G 1–2 for the boiling fowl.

QUANTITIES for 8

8 *chicken joints:* 3 *tsp salt:* 1 *oz cornflour* (3 *Tbs or* 30 *g*): ¼ *tsp pepper.*

Mix the cornflour and seasonings and put them in a paper bag with the chicken joints. Shake together until the chicken is well coated.

2 *oz oil or fat* (4 *Tbs or* 60 *g*)

110

Heat this and fry the chicken until it is brown all over. Transfer the chicken to a casserole.

16 *small onions or shallots, skinned: 4 sticks celery, chopped: 4 carrots, sliced: 4 Tbs tomato paste: ¾ pt chicken stock (1½ c or 4½ dl), mixed with the tomato*

Add these to the casserole, cover and cook gently until the chicken is tender.

To Freeze
Cool the chicken quickly by putting the casserole in ice cold water. Freeze the chicken in the casserole, remove and wrap for storage. To save freezer space the meat may be removed from the bone.

To Use
Return to the casserole and heat in the oven at E 400°F G 6 for 1–2 hrs depending on the quantity being heated. One or two portions may be heated more rapidly in the top of a double boiler.

CHICKEN WITH CHERRIES AND GINGER

COOKING TIME: 1 hr: QUANTITIES for 4–5

TEMPERATURE: E 325°F G 3

3 *lb chicken (1½ kg)*

Cut the chicken in portions reserving the giblets and back to make stock. This will take ½ hr in a pressure cooker. Alternatively use a chicken cube with water.

Use a wide, shallow oven-proof dish and arrange the pieces of chicken close together in a single layer.

2 *oz butter (4 Tbs or 60 g): 1 Tbs honey: ¼ pt stock (½ c or 1½ dl)*

Melt the butter and mix with the honey. Use this to brush the top of the chicken pieces. Pour the stock carefully down the side of the dish.

Bake uncovered until the chicken is browned and tender.

8 *oz can stoned cherries (240 g): 1 oz flaked almonds (30 g): 2 oz crystallised ginger (60 g)*

111

Drain the cherries and arrange among the chicken pieces. Sprinkle the ginger and almonds over the chicken.

To Freeze
Stand the baking dish in cold water to cool the chicken as quickly as possible. Either freeze it in the dish or transfer carefully to a rigid container. If frozen in the dish, dip in warm water to loosen, remove and wrap.

To Use
Transfer the chicken to an oven-proof dish and heat at E 400°F G 6 for 1–1½ hrs or until the chicken is hot.
Serve with rice boiled in chicken stock.

CHICKEN CURRY

Method 1
Use the CURRY recipe, page 97, substituting 6–8 portions of chicken or boiling fowl for the meat. If chicken is used, the cooking time can be reduced to 1–1½ hrs.
Freeze and re-heat as for CURRY.

Method 2

> 1 *pt frozen curry sauce* (½ *l*), *recipe page* 55.

Thaw the sauce over a gentle direct heat or in the top of a double boiler.

> 6 *pieces roasting chicken: oil for frying*

Wash and dry the chicken. Brown it on both sides in a little hot oil. Add to the curry sauce, cover and simmer until the chicken is tender, about 1 hour. Serve with boiled rice.

CHICKEN À LA KING

COOKING TIME: depends on the method used. QUANTITIES for 8–12

1½ *lb cooked chicken* (750 *g*): 6 *Tbs chopped green peppers, fresh or frozen: two* 7½ *oz cans sliced mushrooms in water* (450 *g*): 4 *canned red peppers or pimentos:* 1 *pt chicken stock* (2 *c or* ½ *l*)
112

Remove the flesh from the chicken and cut it into dice. Put it in the refrigerator until required.

Use the carcase to make the chicken stock by covering with water and boiling 2 hours or pressure cook for ½ hour.

Strain.

Chop the green peppers and slice the red peppers. Drain the mushrooms, keeping the liquor to add to the stock.

2 oz butter or margarine (4 Tbs or 60 g)

Heat this and cook the green peppers and the mushrooms in it for about 5 minutes, stirring frequently.

2 oz cornflour (6 Tbs or 60 g): 1 tsp salt

Add to the pan and blend well.

1 pt milk (2 c or ½ l)

Add the stock and milk and cook until it thickens, stirring constantly. Add the red peppers and the chicken and mix well. If some is to be served straight away, simmer until the chicken is well heated.

To Freeze
Stand the pan in cold water to cool quickly. Divide the mixture between two or more containers, leaving a head space.

Seal and freeze.

To Use
Remove the container and heat in a double boiler. Half the recipe will take about 30 minutes, but break up the mixture as it thaws. If the full amount is being heated it will take about an hour, depending on the width of the pan being used. Serve with either cooked rice, mashed potatoes, sweet corn, green beans or peas.

CHICKEN LIVER PÂTÉ

COOKING TIME: 8–10 mins: QUANTITIES for 8 or more

1 lb chicken livers (½ kg)

Wash the livers, dry on kitchen paper and remove any fibres.

113

1 small onion

Skin and slice finely.

2 oz butter (4 Tbs or 60 g)

Heat some of the butter and fry the liver quickly in it for 3–4 minutes. Remove and put aside to cool.

Heat enough more butter to fry the onion until it is tender. Add any remaining butter and allow it to melt. Put aside to cool a little.

½ tsp salt: pinch cayenne pepper: pinch powdered marjoram: pinch ground mace: 3 Tbs marsala or sherry (or more)

While the liver and onions are still warm put them in the electric blender with the other ingredients. Blend until smooth. Pour into one or more rigid containers. Allow to become cold. If no blender is available rub the mixture through a sieve.

To Freeze
Leave a head space, seal and freeze.

To Use
Thaw in the refrigerator and store there until required. Serve with thin toast as an hors d'oeuvre or use for sandwiches.

CHICKEN PIE To serve cold or hot.

COOKING TIME: 1 hr: QUANTITIES for 6: TEMPERATURE: E 425°F G 7 until the pastry is brown, then E 325°F G 3

3 lb oven-ready roasting chicken (1½ kg)

Cut the chicken into joints, four from the legs, two wings and four pieces from the breast. Use the rest of the carcase to make stock. Cover the carcase with cold water, add the giblets and simmer for 2 hrs or pressure cook for ½ hr. Strain. While this is being made, store the chicken joints in the refrigerator.

8 oz ready-made puff pastry (240 g)

If the pastry is frozen thaw it for rolling out.

4 oz bacon (4 rashers or 120 g): 1 hard boiled egg: 1 small onion

Remove the rinds and cut the bacon into strips. Boil the egg. Skin and chop the onion.

> 1 *beaten egg for brushing: salt and pepper*

Use a 1½ pt pie dish. Put in the chicken, onion and bacon. Sprinkle lightly with salt and pepper. Add ¼ pt of the chicken stock (½ c or 1½ dl)

Cut the hard-boiled egg in to eight pieces and arrange in crevices in the top, whites uppermost.

Roll the pastry, cover the pie and make a hole in the centre. Cut leaves of spare pastry and arrange round the hole. Brush with beaten egg and rest in a cold place for ½ hr. Brush again with egg and bake at the higher temperature until the pastry begins to brown, then at the lower temperature. When the pastry is brown enough, cover the top lightly with a piece of grease-proof paper or foil.

To Freeze
Stand the pie dish in cold water to cool it quickly. When it is cold make some extra jelly by heating ½ pt of the chicken stock (1 c or ¼ l) with 1 tsp gelatine. When the gelatine is dissolved, cool the stock. Pour it into the pie using a small plastic or foil funnel. Wrap the pie closely with double foil, chill in the refrigerator and freeze.

To Serve Cold
Thaw 24–36 hrs in the refrigerator or 6–8 hrs at room temperature. When it is thawed store it in the refrigerator until required.

To Serve Hot
Bake in a fairly hot oven for about 1 hr.

CHICKEN WITH TARRAGON SAUCE

COOKING TIME: 1 hr: QUANTITIES for 4

TEMPERATURE: E 325°F G 3

1½ *Tbs oil: 4 portions chicken*

Heat the oil and fry the chicken until golden brown on both sides.

If the pan is suitable for oven cooking, lift the pieces of chicken out. Otherwise transfer them to a shallow oven-proof dish.

½ oz butter (15 g)

Add this to the oil and allow to melt.

2 Tbs cornflour

Add to the pan and mix and blend well.

½ pt milk (1 c or ¼ l)

Add to the pan and stir continuously until it boils.

1 Tbs chopped fresh tarragon: 1 tsp salt: 1 tsp chopped fresh thyme: pinch of pepper: 3 Tbs dry white wine.

Add to the sauce. Either return the chicken to the sauce or pour the sauce over the chicken in the baking dish. Cover with the lid or with foil and cook in the oven until the chicken is tender.

To Freeze

Chill quickly by standing the casserole or pan in cold water. Using a knife and fork, remove the meat from the bones and pack it in a rigid freezer container. Pour the sauce over, seal and freeze.

To Use

Thaw and heat in a double boiler. If liked, the sauce may be thickened further by adding a beaten egg and a little cream. Make sure the mixture is thoroughly heated.

CREAMED CHICKEN

A very useful recipe to freeze. It can be made with chicken cooked specially or with chicken left from a roast or boiled one.

COOKING TIME: depends on method used to make stock and whether chicken is cooked specially.

QUANTITIES for 6–8

1½ pt cubed cooked chicken (¾ l): ¾ pt chicken stock (4½ dl)

Remove chicken from bones and cut in cubes. Put in the refrigerator for use later.

If chicken stock has been made during the original cooking,

use this, otherwise make stock from the carcase by pressure cooking it for ½ hr, with a bay leaf and a piece of carrot and onion. Strain. (Ordinary boiling will take 2 hrs.)

1½ oz chicken fat, butter or margarine (3 Tbs or 45 g): 1½ oz cornflour (4½ Tbs or 45 g): ½ Tbs salt

Melt the fat and add the flour and salt. Cook for a minute or two.

½ pt milk (1 c or ¼ l)

Add to the fat and flour, together with the ¾ pt chicken stock. Stir until it thickens and boils. Add the chicken and simmer for 5 mins.

To Freeze
Cool as quickly as possible.
Divide into containers according to the amount required for one meal. Leave a head space. Seal and freeze.

To Use
Heat in the top of a double boiler, breaking up as it begins to thaw. For half the recipe heat for about 45 mins.
Serve in a border of cooked rice or creamed potatoes with vegetables or salad or use as a filling for patties.
To add colour to the dish sprinkle the potato with chopped parsley and add chopped sweet peppers, raw or canned, to the rice.

DUCK IN ORANGE SAUCE

COOKING TIME: Duck 1½–2 hrs. Sauce 1 hr: QUANTITIES for 4–5

3½–4 lb duck (1½–2 kg)

Roast the duck at E 325°F G 3. Remove from the pan and drain. Carve it into neat portions. If desired, it may be completely removed from the bone but it looks better if the meat is left on the leg and wing bones. Cool the meat rapidly and then store it in the refrigerator while the sauce is made. Use the carcase with 1 pint water to make stock in the pressure cooker, ½ hr cooking. Strain.

½ oz duck dripping (1 Tbs or 15 g): 1 oz onion (30 g): 1 rasher streaky bacon

Skin and slice the onion. Remove the rind and chop the bacon. Heat the dripping and fry onion and bacon until golden.

1 *Tbs flour*

Add to the pan, mix well and cook gently until the flour is yellow. Remove from the heat.

1 *tsp tomato paste:* ¾ *pt duck stock* (1½ *c or* 4½ *dl*): ½ *tsp salt: pinch pepper:* 1 *tsp mixed chopped herbs*

Add to the sauce, mix, return to the heat and simmer for 20 mins.

1 *oz onion* (30 *g*): ½ *oz butter* (1 *Tbs or* 15 *g*)

Skin and chop the onion and cook it in the butter in a separate pan until golden.

¼ *pt red wine* (½ *c or* 1½ *dl*): ½ *bay leaf*

Add to the pan and boil for 2–3 mins.

1 *orange*

Peel the rind thinly and add half the rind and all the juice to the pan, together with the duck sauce. Simmer for 5 mins.

Cut the remaining rind in shreds and blanch it in boiling water for 5 mins.

Strain the sauce.

1 *tsp red currant jelly*

Add to the sauce together with the orange shreds. Heat until the jelly dissolves. Cool quickly.

To Freeze
Put the cold duck in a rigid container and pour the cold sauce over it, seal and freeze.

To Use
Either heat in the top of a double boiler for ½–¾ hr or heat in an uncovered casserole in the oven E400°F G 6 for 45 mins or until well heated.

POTTED DUCK

QUANTITIES for 8 (for two 1 pt or ½ l containers)

8 oz cooked duck (weighed without bones) (250 g): 4 oz duck dripping ($\frac{1}{2}$ c or 120 g): or use half butter and half dripping: 2 Tbs marsala or dry sherry: pinch ground cloves: $\frac{1}{2}$ tsp salt: 2 tsp lemon juice: about $\frac{1}{4}$ pt duck stock or gravy ($\frac{1}{2}$ c or $1\frac{1}{2}$ dl)

Cut the duck meat into small pieces. If you have an electric blender, blend all the ingredients, using enough stock to make a thick, smooth mixture.

Alternatively, mince the duck finely and then pound in a mortar or rub through a sieve to make it smooth.

To Freeze
Pack in small rigid containers, tapping the container sharply to remove air if the mixture has been blended. Seal and freeze.

To Use
Thaw in the refrigerator and store there until required. Use it for hors d'oeuvre, served with thin toast, or use it as a spread for open sandwiches, cocktail snacks, or as a sandwich spread. If desired, add a little pepper where practicable.

Chapter Fourteen

VEGETABLES AND HERBS

In commercial freezing great importance is attached to having the freezing plant located as near to the farm as possible so that there is minimum delay between harvesting and freezing. The variety of vegetables and the method of growing are controlled, tests being made to find the ideal moment for harvesting to give a top quality product.

Those doing home freezing of vegetables are advised to harvest the crops in the early morning and to pick only small amounts such as can be frozen within a short space of time. If it is necessary to keep the vegetables for more than 2–3 hours from picking to freezing they should be stored in the refrigerator meanwhile.

The best stage of maturity at which to freeze vegetables is that at which they would be best for table use.

The freezer owner who does not have garden produce can

freeze prepared vegetable dishes such as vegetable purées, see page 122; potato dishes, see pages 127 and 133; and other recipes, see pages 127–34. If you are able to purchase fresh vegetables in first-class condition and at advantageous prices you may think it worth while to freeze some, for example young broad beans, green beans, tomatoes for cooking, celery for flavouring, herbs, sweet peppers and other seasonable vegetables.

GENERAL PREPARATION

The normal preparation of washing, paring, trimming and cutting is required for all vegetables to be frozen. In addition they need to be blanched, unless they are intended only for flavouring purposes or are going to be stored for only 2 or 3 weeks.

Beyond this time enzymes present in the vegetables will cause loss of quality. The object of blanching is to inactivate the enzymes. Otherwise they cause changes in colour, flavour, texture and nutritive value, especially a loss in vitamin content.

Blanching serves also to soften the vegetables and make close packing easier. It drives air out of the cells and reduces oxidation. It shortens the final cooking time.

Equipment for Blanching

As rapid cooling after blanching is desirable, it is a good plan to make plenty of ice in the freezer before harvesting the vegetables.

A pan to hold eight pints (4 l) of water is needed and a similar sized vessel for iced-water for cooling.

For holding the vegetables you will need either a wire chip basket, a colander or muslin.

Blanching

After the normal washing and preparation the vegetables should be divided into 1 lb ($\frac{1}{2}$ kg) lots ($\frac{1}{2}$ lb or 240 g if green leafy).

Bring 8 pints (4 l) of water to the boil in the pan and put one portion of vegetables in the wire basket or alternative, lower into the boiling water, keeping full heat under the pan. Bring the water back to the boil and count the blanching time from then. For times see notes on individual vegetables, pages 121–8. Make sure the vegetables have room to move about freely.

120

Lift out the vegetables and plunge into ice cold water. Shake the container frequently and make sure the vegetables are not lumping together. Before removing them from the water break open one piece to test to see if it is cold.

Drain thoroughly.

Packing and Freezing

Pack the vegetables closely in containers, bags or boxes. Chill them in the refrigerator before freezing.

With certain vegetables, for example peas, corn and beans, it is an advantage to have them "free-running" instead of clumped together in the packet. Freeze the vegetables spread out on trays and then pack loose in bags. It is then possible to remove only as much as is required for immediate use.

Storage should be at $-18°C$ ($0°F$) or below, with the minimum of fluctuation in temperature. Vegetables which have been kept at higher temperatures for any length of time lose their colour and finally develop disagreeable flavours.

COOKING

Because of the blanching process prior to freezing, vegetables need only $\frac{1}{2}-\frac{2}{3}$ the normal cooking time. Use the minimum amount of water, $\frac{1}{4}-\frac{1}{2}$ pt ($1\frac{1}{2}-2\frac{1}{2}$ dl) boiling salted water per 1 lb ($\frac{1}{2}$ kg) of vegetables. Whenever possible vegetables should be cooked while still frozen. Exceptions to this are corn-on-the-cob; whole green peppers; and vegetables where the pieces need to be separated for quick cooking (allow partial thawing). Cooking times will be found with the individual vegetables in the information which follows.

Cooking frozen vegetables in fat

This can be done either in a casserole in the oven or in a pan on top of the cooker. Use about $\frac{1}{2}$ oz (1 Tbs or 15 g) of butter or margarine per lb ($\frac{1}{2}$ kg) of vegetables. Heat the fat, add the vegetables, and cook until the pieces can be separated. Cover and continue to cook until tender, time depending on the temperature of the oven.

Using frozen vegetables in recipes

They can be substituted for fresh vegetables in any recipe, except where raw salad vegetables are required.

VEGETABLE PURÉES

These are very useful to have for soups, sauces, infant feeding and for special diets. Vegetables are cooked in the usual way, rubbed through a sieve or put in the electric blender and then packed in small containers, chilled and frozen. Tap the container several times on the edge of the bench to remove bubbles of air, specially important if the blender has been used.

For single portions for infant feeding and special diets, freeze the purée in ice trays, remove the cubes and pack in bags.

To heat the purée put it in a double boiler, in a basin over boiling water, or in a covered dish in the oven.

Alphabetical List of Vegetables

ASPARAGUS

Preparation Do not cut it until you are ready to freeze it as asparagus loses flavour very rapidly. Trim to lengths to fit the container, using one which will hold only the edible length. It is a waste of freezer space to include the woody bits. Cut off any scales or bracts and wash the stalks very thoroughly. Sort according to thickness, but do not tie in bundles.

Blanching Pieces ¾–1 inch (2–2½ cm) for 4 mins; ⅜–¾ inch (1–2 cm) for 3 mins; ⅜ inch (1 cm) for 2 mins. Shake the container gently all the time to keep the stalks moving gently for even heating. Chill for 10–15 mins.

Drain and pack alternate tips to tails to make a good close pack.

To Use Boil 5–8 mins. Serve hot with melted butter or cold with French dressing.

BEANS, BROAD

Preparation Shell and sort according to size. Freeze only young ones.

Blanching Very small young beans 2 mins; medium size 3 mins; large 4 mins. Chill for the same times.

Freezing Pack closely in bags or boxes.

To Use Boil 12–15 mins. Thicken the cooking liquid with roux and add chopped parsley or other herbs.

BEANS, FRENCH

Preparation Use only young tender beans which snap when bent. Wash and cut off the ends.

Blanching 3 mins and chill 3 mins.

Freezing Pack in bags or boxes.

To Use Boil 12–15 mins. Dress with melted butter.

BEANS, RUNNER

Preparation They can be sliced finely but will have a better flavour if trimmed and cut in 1 inch (2½ cm) lengths.

Blanching 2 mins and chill 2 mins.

Freezing Pack in bags or boxes.

To Use Boil 12–15 mins. Dress with melted butter.

BEETROOT

Preparation Only small young beetroot should be frozen whole. Twist off the tops leaving about 2 ins (5 cm) still attached. Wash gently and blanch about 20 mins. Chill 20 mins, skin. Larger ones cook until tender.
Chill and remove the skins. Dice or slice.

Freezing Pack dry in boxes, bags, or in a sauce.

To Use Thaw dry ones at room temperature, dress with lemon or vinegar or use as a garnish for salads. In a sauce, heat in a double boiler and use as a vegetable.

BROCCOLI, GREEN SPROUTING

Preparation Soak for 30 mins in cold salted water to bring out any insects. Rinse very thoroughly and discard any woody stems. Trim the remainder to an even size.

Blanching Thin stalks 3 mins; medium 4 mins; thick 5 mins. Chill for the same times. Drain well.

Freezing Pack in boxes with the heads and stalks alternating.

To Use Thaw at room temperature and use raw in salads. Boil 5–7 mins. Serve with melted butter or with Béchamel sauce. Add grated cheese with either method and brown under the grill.

BRUSSELS SPROUTS

Preparation Soak for 30 mins in cold salted water. Sort to even sizes.

Blanching Small 3 mins; medium 4 mins; large 5 mins. Chill for 6–8 mins.

Freezing Pack in bags or boxes.

To Use Boil 3–4 mins. Toss in melted butter or serve with Béchamel sauce, plain or with added cheese.

CABBAGE

As this is in season all the year round it is a waste of freezer space to store it except for a cooked dish, see recipes pages 129 and 130.

CARROTS

Preparation There is no point in freezing these except to prolong the life of very young and tender carrots. Wash and cut off the tops, freeze whole, or in strips or dice.

Blanching 5 mins and chill 5 mins. Rub off the skins.

Freezing Pack in boxes top to tail for whole ones.

Cooking Boil 5–10 mins. Serve with melted butter and chopped parsley. See also Vichy carrots, page 130.

CAULIFLOWER

Preparation Use good quality compact heads. Break into sprigs about 1 inch (2½ cm) across. Wash and drain.

Blanching for 3 mins and chill for 4–5 mins.

Freezing Pack closely in bags or boxes.

To Use Boil 5–8 mins. Thicken the liquid with roux and add chopped parsley, or grated cheese. Or sprinkle the drained cauliflower with grated cheese, dot with butter and brown under grill.

CELERY

Preparation Not satisfactory for serving raw but can be used for flavouring or as a cooked vegetable. Wash, trim and cut in 1 inch ($2\frac{1}{2}$ cm) lengths.

Blanching not necessary. If to be used as a vegetable, boil gently until almost tender. Chill quickly by standing the pan in cold water.

Freezing Unblanched pieces for flavouring, store in bags. Cooked celery, pack in boxes, covered with cooking liquid.

To Use For flavouring add to soups, stews, etc, without thawing. Cooked celery heat in a double boiler or in a pan in the oven. Thicken the liquid for a sauce. See Braised Celery, page 131.

CORN-ON-THE-COB

Preparation Use only small, young, still milky cobs and process as soon as possible after gathering. Remove husk and silk and wash the cobs. Grade for size.

Blanching 4–5 mins. Chill 15 mins.

Freezing Wrap each cob separately in Cellophane and pack in a carton or bag. Alternatively cut the corn off close to the cob and pack it in small containers.

To Use Boil on the cob for 5–10 mins; corn only, 3–4 mins. Serve with melted butter or add Béchamel sauce to the loose corn.

COURGETTES See recipe pages 131 and 134.

MUSHROOMS

Preparation Use button mushrooms. Wash and dry. Very small ones of less than 1 inch (2½ cm) diameter can be left whole. Quarter or slice larger ones.

Blanching not necessary.

Freezing Pack in bags or boxes.

To Use No need to thaw. Use in any recipe, including frying.

ONIONS

Useful peeled, chopped and in small packets. For short storage periods only or they will lose flavour. No blanching is needed. Use while frozen.

PARSNIPS

Preparation Peel and slice ½ inch (1 cm) thick.

Blanch 2 mins and chill 2 mins.

Freezing in bags or boxes.

To Use Boil or cook in fat, see page 121.

PEAS

Preparation Use only young tender peas. Shell.

Blanching 1 min. and chill 5 mins.

Freezing If required loose, freeze on trays before packing in bags.

To Use Boil 6–8 mins and dress with butter or cook in butter, see page 121.

PEPPERS, SWEET

Preparation Choose fresh, firm ones. Wash, remove stem and seeds. Slice, cut in pieces or dice.

Blanching not necessary.

Freezing Put in bags so that a small amount can be removed as required.

To Use Add to soups and casseroles without thawing. They can be chopped while still frozen. Thaw sliced ones and use in salad or hors d'oeuvre.

POTATOES

Preparation Raw potatoes are not satisfactory but cooked ones may be frozen. I think it is a waste of freezer space to do this except possibly Duchess potatoes for garnishing. You may think it worth while to freeze chips to avoid last minute frying or cut down the number of frying times. Roast potatoes take as long to thaw as to cook in the first place but results are good.

Duchess Potatoes See page 133.

Chips
Fry in the usual way. Drain very thoroughly on absorbent paper. Cool quickly and pack in bags or boxes.

To Use Spread on a baking tray and heat at E 400°F G 6 for 5–10 mins. Test one to see if properly thawed and hot.

Roast Potatoes
Roast in the usual way until they are lightly browned. Cool as quickly as possible. Pack in bags.

To Use Put in a baking tin and thaw and heat at E 400°F G 6 for $\frac{1}{2}$–$\frac{3}{4}$ hr depending on size. Test one to see if it is hot. Salt and serve.

SPINACH

Preparation It should be very freshly gathered and it is not wise to try to process more than $\frac{1}{2}$–1 lb (250–500 g) at a time. Pick over and wash in running water. Drain.

Blanching Use 2 gallons (8 l) of boiling water to 1 lb (500 g). Agitate for 2 mins and chill for 2 mins.

Freezing Pack tightly in bags or boxes.

To Use Boil 4–6 minutes with little or no water. Drain, chop and dress with butter. Or sieve and mix with thick cream.

TOMATOES
Not satisfactory for serving raw.

For Frying and Grilling
Choose good quality firm tomatoes and wrap individually in Cellophane. Pack in boxes. Thaw enough to cut in half. Fry or grill.

For Purée
Blanch 2 mins, chill and skin. For a raw purée, chop and pack with a little head space.

For a cooked purée there is no need to skin, cook to a pulp and sieve. Pack as before.

To Use Add raw purée to soups, stews and sauces, without thawing. Cooked purée thaw or use frozen as required.

Tomato Juice made from raw purée, strained, may be frozen for use as a breakfast drink or for tomato juice cocktails. Thaw and add flavouring to taste. Serve chilled. For cooked juice see page 137.

TURNIPS

Preparation Use only small young ones for flavouring purposes. Peel and leave whole or cut in pieces.

Blanching for 2 mins. Chill 2 mins.

Freezing in bags or boxes.

To Use Add without thawing to soups, stews, etc.

HERBS

Preparation Wash, drain and dry thoroughly. Do not chop. They are more satisfactory left whole and chopped while still frozen, or partly thawed. They are usually only suitable for cooking.

Blanching is not necessary.

Freezing Pack closely in small bags or boxes. Parsley may be frozen in a tight ball and then grated frozen.

To Use Add to sauces, stews, soups etc. See also recipes for Mint sauce, page 58.

Herb Butters, pages, 61–2.

FREEZING COOKED VEGETABLE DISHES

From the health point of view this is not a good thing to do because the double heating process involved causes loss of nutrients, particularly vitamin C.

It is, however, convenient sometimes to do this and I have included a few recipes I find are satisfactory so far as texture and palatability are concerned.

CREAMED CABBAGE

COOKING TIME: 15 mins: QUANTITIES for 8–10

3 *lb cabbage* (1½ *kg*): ½ *pt water* (1 *c or* ¼ *l*): 2 *tsp salt*

Remove any tough outer leaves from the cabbage. Cut in half and cut out the thick stalk. Wash and drain the cabbage and cut it in ¼ inch wide strips.

Boil the water and salt. Add the cabbage, cover and boil for 5 mins.

1 *oz margarine or dripping* (2 *Tbs or* 30 *g*)

Add to the cabbage.

2 *oz flour* (6 *Tbs or* 60 *g*): 2 *oz grated cheese* (½ *c or* 60 *g*): ½ *pt water* (1 *c or* ¼ *l*)

Blend the flour and cheese to a smooth cream with the water. Add it to the cabbage, stir well, and bring to the boil. Cook for another 10 mins, or until barely tender.

To Freeze
Stand the pan in cold water to cool as quickly as possible. Transfer to rigid containers, leaving a head space. Seal and freeze.

To Use

Thaw in a double boiler or over a very gentle heat, stirring frequently to prevent sticking. Bring to the boil and serve. It is very good served with pork, veal or sausages.

RED CABBAGE

COOKING TIME: ½–¾ hr: QUANTITIES for 4

1 *lb red cabbage (½ kg): 1 large apple: 1 large onion*

Wash the cabbage, cut it in quarters and remove the hard centre stalk. Skin the onion and peel and core the apple. Put all these in a shredder to give slices about ¼ inch thick (½ cm). Put in a pan or casserole.

1 *oz dripping or lard (2 Tbs or 30 g): 2 Tbs water: 2 Tbs vinegar*
1 *tsp salt: pinch pepper: 1 Tbs brown sugar*

Add these to the cabbage, cover and boil gently until the cabbage is tender. There should be hardly any liquid left when the cooking is finished.

To Freeze

Chill quickly by standing the pan in cold water. Pack the cabbage into a rigid container, leaving a little head space. Cover and freeze.

To Use

Thaw and heat either in a moderate oven or in a double boiler or over a direct gentle heat. Use a fork to break it up as it begins to thaw.

VICHY CARROTS

COOKING TIME: 10–15 mins: QUANTITIES for 4

1 *lb frozen sliced or small whole carrots (½ kg): 1 oz butter (2 Tbs or 30 g): 1 tsp sugar: ½ tsp salt*

Melt the butter in a saucepan or casserole and add the other ingredients. Cover and cook until the carrots are tender.

1 *Tbs chopped parsley*

Serve sprinkled with the parsley.

BRAISED CELERY

2 large head of celery: 2 carrots: 2 small turnips: 2 onions: 4 rashers bacon: sprig parsley: piece of bay leaf: stock

Separate the celery stalks and wash them well. Cut in lengths to fit an oblong freezer box.

Peel and slice the vegetables.

Remove the rind and chop the bacon.

Wash parsley and bay leaf and make the stock (a chicken cube will do).

Put the carrot, turnip, onion and bacon in the bottom of a large stewpan. Add the herbs and just enough stock to cover. Arrange the celery on top of the vegetables, cover and cook gently until the celery is just tender.

Lift the celery out and divide it between two or more containers.

3 Tbs potato starch

Measure the stock left in the pan and make it up to 1 pt (2 c or ½ l). Blend the starch with a little cold water. Return the stock to the pan, add the starch and stir until it thickens. Pour this over the celery, together with the chopped vegetables and bacon.

To Freeze
Make sure there is a head space left in each container. Cool as quickly as possible, seal and freeze.

To Use
Empty into a flat baking dish and heat, uncovered at E 400°F G 6 for 45 mins. Stir before serving.

COURGETTES WITH TOMATOES

2–3 lb courgettes (1–1½ kg): 1 lb tomatoes (½ kg): 1 large onion

Wash the courgettes and remove any stalk. Cut them in slices about 1 inch thick (½ cm). Wash and cut up the tomatoes. Skin and chop the onion.

1 *oz fat (2 Tbs or 30 g)*

Heat the fat in a stewpan and fry the tomatoes and onions for about 5 minutes. Add the courgettes.

½ *tsp salt: pinch of mixed herbs: 1 tsp sugar: ½ pt stock (1 c or ¼ l)*

Add to the pan, cover and cook gently until the courgettes are beginning to become tender but are not completely cooked. They will cook some more during thawing and heating.

To Freeze
Cool as quickly as possible by standing the pan in iced water. Put the courgettes and sauce in a rigid container, leaving head space. Seal and freeze.

To Use
Thaw and finish cooking either over a direct gentle heat, in a double boiler or in the oven. Time 30 mins or ¾–1 hr in the oven.

Chopped parsley: salt and pepper

Add more seasoning if necessary and serve sprinkled with chopped parsley.

TOASTED PEPPERS

COOKING TIME: 10 mins: QUANTITIES for 8 or more

12 *large green peppers*

Wash and dry the peppers. Put them on a grill rack, at least 6 inches from the heat so that they will cook without burning. Grill about 10 minutes or until they are brown on both sides. Turn frequently.

When the peppers are soft, allow them to cool.

Peel off the skin, cut out the stem and remove the seeds.

Cut the peppers lengthwise in 2 inch (5 cm) strips. Place in a deep dish.

¼ *pt olive oil (½ c or 1½ dl): 2 Tbs wine vinegar: pinch of garlic salt or plain salt to taste*

Mix these together and pour over the peppers. Leave until quite cold.

To Freeze
Pack in rigid containers, leaving a head space. Seal and freeze.

To Serve
Either thaw at room temperature and serve cold with meat or fish or heat at E 400°F G 6 for 30 mins. Alternatively, heat in a double boiler for 30 mins.

DUCHESS POTATOES

COOKING TIME: about 30 mins before freezing: QUANTITIES for 8

2 lb old potatoes (1 kg)

Boil the potatoes until tender and sieve while still hot. Return to the pan and dry them over a gentle heat or in the oven.

2 oz butter or margarine (60 g): *2 eggs or 4 yolks*

Beat in the fat and then the egg.

Hot cream or milk

If necessary, add a little to make the mixture soft enough for piping.

Pinch of grated nutmeg or mace: salt and pepper

Season to taste but be sparing with the pepper.

Put into a forcing bag with a large rosette nozzle and pipe in pyramids or other shapes, on to Cellophane paper on a tray.

To Freeze
Freeze uncovered until firm, remove from the tray, pack and seal.

To Use
Place on greased trays and put in a cold oven set at E 400°F G 6 and cook for 20 mins or until heated and lightly browned. Before baking they may be brushed with egg wash or melted butter. A pre-heated oven may be used but the potatoes are then inclined to stick to the tray. Non-stick lining paper will prevent this.

RATATOUILLE

A useful dish to make to extend the season of these imported vegetables.

8 oz green or red sweet peppers (240 g): 8 oz courgettes (240 g): 8 oz aubergines (240 g): 1 large onion: 12 oz tomatoes (360 g)

Wash and remove seeds from the peppers. Slice them.
Wash the courgettes and slice them thickly.
Wash the aubergines and remove the stalk and tip, slice.
Peel and slice the onion. Skin and slice the tomatoes.

$\frac{1}{4}$–$\frac{1}{2}$ pt olive oil ($\frac{1}{2}$–1 c or $1\frac{1}{2}$–$2\frac{1}{2}$ dl): pinch dried garlic: pinch of thyme: piece of bay leaf

Heat some oil in a frying pan and fry the peppers, onion, herbs and garlic until they are beginning to soften but are not completely cooked. Put them in a foil or foil-lined baking dish.

Heat some more oil and fry the aubergines and courgettes until they are brown. Arrange on top of the other vegetables. Toss the tomatoes in the pan for a minute or two and arrange them among the vegetables.

Salt: 4 Tbs white wine

Season lightly with salt and add the wine.

To Freeze

Leave to become cold, chill in the refrigerator and then seal and freeze. If in a foil-lined baking dish, remove and re-wrap.

To Use

Transfer to the baking dish or put the foil container in the oven, uncovered. Cook at E 325°F G 3 for $1\frac{1}{2}$ hrs or at E 400°F G 6 for 1 hr or until the vegetables are completely cooked and the dish hot.

Chapter Fifteen

FRUIT

A limited range of frozen fruits have been available from the commercial freezers for a number of years. These are generally of very good quality and techniques of freezing the more delicate types such as strawberries have been improved.

If you grow your own fruit, freezing is the easiest method of preserving the surplus. If you buy fruit, the price does not usually make it worth while freezing raw fruit, but it is worth while freezing cooked fruit in the form of purées, sauces and sweets, to have as convenience foods. Also, if you have a favourite fruit which is not readily available from the commercial freezer, it is worth while freezing some for yourself.

Not all fruits are satisfactory frozen raw but most can be cooked and frozen. Raw soft fruits generally give the best results, especially raspberries and currants. Some fruits which discolour easily, like peaches, apricots and apples, need special treatment if they are to be frozen raw.

Only prime quality fruit should be frozen whole, the rest should be cooked before freezing. This is a very good way of using windfalls and damaged fruit.

PREPARATION

On exposure to air enzymes in certain fruits cause colour changes, chiefly browning. Blanching to inactivate the enzymes can be used if the fruit is going to be served cooked but some other method is needed if the fruit is to be served raw.

The simplest method of dealing with this problem is to handle only small amounts of the difficult fruits (apples, peaches, apricots and pears) at a time, to cover them with syrup and to get them into the freezer quickly.

Alternative treatments are to use a salt solution, or a lemon solution to preserve the colour during preparation. Or you can add an ascorbic acid (vitamin C) solution to the fruit before packing or to the syrup used for covering the fruit during freezing.

PROPORTIONS
Salt Solution: 1 Tbs. salt to 2 qts. (2 l) of water. Put the fruit into this as it is prepared and leave for a maximum of 15–20 minutes. Drain, rinse and blanch or pack in syrup.

Lemon Juice Solution: 2 Tbs. lemon juice to 2 qts. (2 l) water. Put the fruit into this as it is prepared and leave for 1 minute. Drain and pack in syrup.

Ascorbic Acid Solution: Buy 500 or 1,000 mg. tablets from the chemist. Crush the tablets and dissolve them in the syrup just before using it. For every qt (l) of syrup use 3,750 mg. of ascorbic acid. Use enough syrup to cover the fruit. For dry sugar packs use 375 mg. dissolved in 1–2 Tbs. cold water and sprinkle this over the fruit at the rate of 375 mg. per pound ($\frac{1}{2}$ kg).

ADDING SYRUPS AND SUGAR

Using Dry Sugar: 1 lb sugar ($\frac{1}{2}$ kg) to 4 lb fruit (2 kg) is usually sufficient. Spread the prepared fruit in shallow layers, sprinkle the sugar over it and wait for the juice to run before packing it. If only small amounts of fruit are being packed in rigid container ($\frac{1}{2}$ lb or $\frac{1}{4}$ kg), the sugar can be sprinkled over the fruit in the container. This avoids handling the fruit when it is partially softened by the sugar.

Using Syrup: Allow $\frac{1}{2}$ pt ($\frac{1}{4}$ l) syrup per lb ($\frac{1}{2}$ kg) fruit. Syrup must be cold when it is added to the fruit. A light syrup can be made by mixing sugar into cold water (4 oz or 120 g to 1 pt or $\frac{1}{2}$ l). A heavy syrup usually needs to be heated to dissolve the sugar. Chill before using.

For a heavy syrup use from 8 oz to 1 lb sugar ($\frac{1}{4}$–$\frac{1}{2}$ kg) to 1 pt ($\frac{1}{2}$ l).

PACKING

Tubs and rigid containers are the best to use when syrup has been added. To keep the fruit submerged, put a little crumpled waxed or greaseproof paper on top before placing the lid on.

With dry fruit there is no need to leave a head space but leave $\frac{3}{4}$–$\frac{1}{2}$ inch ($\frac{1}{2}$–1 cm) with syrup.

THAWING

Fruit to be served raw is best thawed slowly in the refrigerator but for quicker thawing it may be emptied into a bowl and thawed at room temperature. For even faster thawing stand the container in cold water for $\frac{1}{2}$–1 hour.

Serve raw fruit while it is still frosty.

Approximate thawing times: 1 lb ($\frac{1}{2}$ kg) fruit needs 3–4 hrs at room temperature: 8–10 hrs in the refrigerator: $\frac{1}{2}$–1 water.

Fruit for cooking can be cooked either frozen or tha better to partially thaw fruit which will be used in a ta helps to prevent the pastry from becoming soggy. Thaw e. separate the pieces.

If there is any left-over thawed fruit bring it quickly to the boil. This prevents loss of flavour and colour.

FREEZING FRUIT JUICES

These are very useful to have for drinks, sauces and cold sweets. For full flavoured juices the fruit should be ripe. Freeze the juice in rigid containers allowing one inch head space.

Preparation

Citrus fruit: squeeze the juice, strain and freeze.

Apple juice: use a juice extractor or cook the unpeeled, cut fruit in $\frac{1}{2}$ pt ($\frac{1}{4}$ l) water to 2 lb (1 kg) apples. Cook to a pulp, strain, chill, sweeten to taste, 2 lb apples gives approximately $\frac{3}{4}$ pt juice.

Berries, cherries, grapes, currants: crush the fruit and heat slowly to simmering point, stirring frequently. Strain the juice, sweeten if desired, chill and freeze.

Strawberries: crush the fruit and strain without heating.

Plums: add water to cover and simmer until soft. Strain, sweeten to taste, chill and freeze.

Tomatoes: heat cut-up tomatoes until the juice just boils. Sieve, cool, add salt to taste, chill and freeze. Raw juice, see page 128.

To Use

Thaw in the refrigerator for 12 hrs per pint ($\frac{1}{2}$ l) container: or at room temperature for 2–3 hrs; or with the sealed container in cold water for about 45 mins.

FREEZING FRUIT PULP AND PURÉES

This is a useful way of preserving very ripe or damaged fruit.

Preparation

Wash and prepare the fruit as for cooking. For a raw purée mash and pulp the fruit or rub through a sieve. For minimum cooking

137

heat the fruit until the juice flows and then mash or sieve. For fruits like apples, cook until tender before pulping or sieving. Cook apricots and peaches 2–3 mins to preserve the colour.

An electric blender can be used to make the purée but use it for the minimum time to avoid making the purée frothy. Tap the containers sharply on the table to knock out air before sealing.

Sweeten if necessary.

To Use
Thaw in the refrigerator for 12 hours per pt container ($\frac{1}{2}$ l); 2–3 hrs at room temperature; 45 mins with the sealed container in cold water. If the purée is to be used hot it may be heated without thawing, preferably in a double boiler.

Use for sweets, infant foods, special diets, sauces and jams.

FRUIT SALAD

Any fruit suitable for freezing may be used, including grapes. Especially good is a mixture of early summer fruits such as cherries, raspberries and currants. Prepare according to the type of fruit and freeze in syrup. Thaw in the refrigerator or at room temperature in the unopened container. Serve while still chilled.

It is also very satisfactory to use small amounts of frozen fruit in fresh or canned fruit salad. Make the salad and then 2–3 hrs before serving tip in the frozen fruit. It will make the salad beautifully cool and the fruit should be thawed by the time it is required. This is a good way of using a few strawberries or other berries.

WHOLE FRUIT FOR DESSERT

This is a suitable method for strawberries, peaches and plums and other dessert fruit. Use only perfect, very firm but ripe fruit. Wash and drain. Put strawberries straight into rigid containers but first wrap other fruits singly in Cellophane.

To Use Thaw in the sealed container in the refrigerator and do not open the container until the fruit is to be served, when it is only just thawed but still frosty.

APPLES

I find the most useful way of storing these is as a dry pack to use for pies and puddings, as purée or sauce, or made up in pies, puddings and sweets.

Dry Pack

Preparation: Use firm apples. Peel, core and slice fairly thickly, cutting an average-size apple in to 16 pieces. As apples brown quickly on exposure to air, work fast with small quantities and slice them straight into salted water, see page 135. Then drain and pack at once into polythene bags, seal and freeze.

As an extra insurance against discoloration and to enable them to be packed very tightly, after peeling blanch them in plenty of boiling water until they are just pliable, about 2 minutes. Plunge at once in ice water to cool. Drain and pack.

With either method the apples can be packed with or without sugar. Sprinkle the sugar between the layers during packing, the amount according to taste. They may be packed in syrup but to me this seems a pointless waste of freezer space.

To Use: For pies and tarts thaw enough to separate the slices. For stewing and similar purposes cook while still frozen or partially thawed. Use in recipes in the same way as fresh apples.

Apple Purée or Pulp

This method is very useful and the apples take up less freezer space than when in slices. Damaged apples can be used provided the bad parts are cut away. It is not necessary to peel the apples unless you particularly want a very light-coloured purée. If you do not want to sieve the apples it is necessary to peel and core before cooking. Cut in pieces and stew with the minimum amount of water needed to prevent sticking. When quite soft, mash to a pulp, put in the blender or sieve. Sweeten or leave without sugar. Pour into rigid containers or polythene bags standing in a carton, see page 27. Leave a head space. Chill and freeze.

To Use Heat without thawing or thaw in the container in the refrigerator or at room temperature. It may be used in any recipe in place of fresh apple pulp or purée or use for apple sauce. For some recipes see pages 162–3.

APRICOTS

Unless you are lucky enough to be able to get apricots picked fully ripe there is not much point in freezing them because the flavour will not be as good as canned apricots.

What is useful is to make ices and sweets with canned apricots and freeze them.

To save freezer space raw apricots should be halved and stoned.

Syrup Pack Put the apricots into the cold syrup as they are prepared. Put crumpled waxed or greaseproof paper on top to keep them submerged. Seal and freeze.

To Use Poach gently in the syrup or use in any recipe requiring fresh or cooked apricots.

Dry Pack (to serve raw) To prevent browning it is necessary to treat the apricots with ascorbic acid solution, see pages 135–6. The fruit may be halved or sliced before being sprinkled with sugar and ascorbic acid solution. Pack in bags or boxes and freeze at once.

To Use To serve raw thaw in the covered container in the refrigerator or at room temperature. They may be cooked while frozen or partially thawed and used in any recipe in place of fresh apricots.

Dry Pack (for pies and cooking).
Prepare as above but instead of using ascorbic acid blanch the fruit for 2 minutes in plenty of boiling water, plunge in ice water to cool, drain and pack as before.

To Use Cook while frozen or partially thawed and use in any recipe in place of fresh apricots. For pies and tarts thaw enough to separate the pieces.

Pulp or Purée See page 137.

Whole fruit for Dessert Page 138.

BANANAS

These are satisfactory only when made into a purée for ices and

cold sweets to be frozen, or used in cooked dishes to be frozen.
See recipes pages 150, 152, 163–4.

BILBERRIES (Whortleberries).
Mix the prepared berries with dry sugar and stir to coat well.
Pack and freeze. Can also be frozen without sugar.

To Use For cooking use frozen or partially thawed. Otherwise
thaw in the covered container in the refrigerator or at room
temperature. For quicker thawing empty into a bowl.

BLACKBERRIES
Pick over, discarding imperfect ones. Remove hulls. Wash only if
absolutely necessary and then drain very thoroughly. Mix with
caster sugar, turning to coat well. Pack, seal and freeze. Can also
be frozen without sugar.

To Use Thaw as for bilberries and use in place of raw fruit in any
recipe. For cooking there is no need to thaw.

Purée and Pulp See page 137.

BLACK, RED OR WHITE CURRANTS
Wash on the stalks and drain well. Remove from the stalks discarding any over- or under-ripe currants. Mix with caster sugar
and turn to coat well. Pack and freeze. Can also be frozen without sugar.

To Use for pies and tarts partially thaw. For stewing and similar
uses such as crumble, cook while frozen.

Purée See page 137.

BOYSENBERRIES See Blackberries.

CHERRIES
Sour cooking cherries are very satisfactory frozen dry without
sugar or in a syrup. Sweet red cherries in a syrup are also very
good, much better than most canned cherries.

Dry Pack Remove stalks, wash and drain the cherries. Leave the pips in if you want the cherries for pies and tarts, otherwise remove. Mix the cherries with caster sugar, pack and freeze. For morello cherries add 1 lb ($\frac{1}{2}$ kg) sugar per 5 lb ($2\frac{1}{2}$ kg) fruit.

To Use For pies and tarts partially thaw in the unopened container. For other cooking use frozen or partially thawed.

In Syrup Leave stones in or remove as desired. Cover with cold syrup, leaving a head space. Seal and freeze.

To Use For serving sweet cherries raw, thaw in the unopened container. For stewing cook gently from the frozen state.

CHESTNUTS

Wash nuts. Cover with water, bring to the boil, drain and peel. Pack in bags or boxes. Use in place of raw chestnuts in any recipe.

They can also be cooked and frozen as a purée for making soups or for sweets, see recipes page 150.

CITRUS FRUIT

As grapefruit, oranges and lemons are available all the year round there is not much point in freezing them except for convenience. Ready prepared in fruit salads and sweets is the most practical way. You may like to preserve some of the more seasonal citrus fruits such as tangerines and mandarins.

Dry Pack Blanch the whole fruit for 2 minutes in plenty of boiling water. This loosens the skin. Plunge in cold water. Peel and break into sections, removing pith and pips. Large fruits like grapefruit which have coarse membranes round the segments are better if the membrane is removed. Pack the segments with sugar sprinkled between the layers. Seal and freeze.

Alternatively the fruit may be sliced across to give thin rounds. Pack as before.

To Use Thaw in the unopened container and use for fruit cocktails, breakfast fruit or any recipe where fresh citrus fruit is required.

Juices See page 137.

Whole Fruit for Dessert See page 138.

COCONUT

Fresh grated coconut freezes satisfactorily. Pack into small cartons or bags. Mix some with coconut milk prior to freezing and use this in curries, sauces, and icings. It can also be frozen in pieces.

CRANBERRIES

I find those purchased from the greengrocer or store, when in season, freeze very satisfactorily.

Dry Pack Pick over, carefully discarding any of poor quality. Pack in boxes and freeze plain or with sugar sprinkled over.

To Use For pies and tarts partially thaw. For other cooking use frozen or partially thawed. See recipes pages 55 and 179.
   Use in any recipe in place of fresh berries.

DAMSONS

Remove stalks, wipe off bloom, rinse and freeze as plums.

FIGS (fresh).

These should be fully ripe and with a tender skin. Peeling is not necessary. Leave whole, halve or slice. Pack dry without sugar or in a syrup adding the juice of 1 large or 2 small lemons to every pint ($\frac{1}{2}$ l) of syrup.

To Use Thaw dry packs in the unopened carton and use as fresh figs. Otherwise cook gently in syrup.

Raw for Dessert See page 138.

GOOSEBERRIES

Both ripe and green gooseberries are suitable for freezing.

Dry Pack Wash, drain, top and tail. Pack in bags or boxes either with or without sugar.

To Use Ripe fruit should be thawed in the unopened container and served while it is still frosty. Alternatively put it in a bowl, pour syrup over and leave it to thaw. Serve as a fruit salad. For pies and tarts thaw completely.

In Syrup Prepare as before, cover with cold syrup leaving a head space, seal and freeze.

To Use Raw ones thaw in the unopened containers and serve as fruit salad. Otherwise cook gently in the syrup from the frozen or partially thawed state.

Purée and Pulp See page 137.

GRAPES
If seedless, leave whole. Cut in half to remove seeds from others.

Dry Pack in small containers.

In Syrup Put in small containers and cover with syrup, leaving a head space. Seal and freeze.

To Use Suitable for cooking or for fruit salad. Add frozen or thawed to fruit salad or thaw at room temperature.

GRAPEFRUIT See Citrus Fruit.

GREENGAGES See Plums.

LEMONS See Citrus Fruit. Also suitable for freezing dry in
wedges or slices. Use as soon as thawed.

LOGANBERRIES See Blackberries.

MANDARIN ORANGES See Citrus Fruit.

MELON
The yellow or orange kinds are the most satisfactory. Use only fully ripe ones of good flavour.

Dry Pack Peel and remove seeds. Cut in slices or cubes. Pack in layers with or without sugar. Alternatively cut in portions for serving, remove seeds. Wrap pieces individually in Cellophane and then over-wrap with freezer paper or a bag.

To Use Thaw in the original container or wrapper in the refrigerator. Use in place of fresh melon, serving while it is still frosty. When it completely thaws a lot of liquid runs out but it is still quite suitable for a salad.

In Syrup Peel and remove seeds. Cut in slices or dice and cover at once with cold syrup, leaving a head space. A syrup made with honey in place of sugar is very good with melon.

To Use Thaw in the unopened container and use while still very cold. It is very useful to add to fresh fruit salads.

MULBERRIES as Blackberries. Freeze firm ones only.

NUTS
Freezing prolongs the life of shelled nuts as it helps to prevent rancidity. See also Chestnuts and Coconut.

ORANGES See Citrus Fruit.

PEACHES
Fully ripe yellow peaches are generally considered the best for freezing but white peaches of good flavour are also suitable.

In Syrup The colour is better if they are peeled without a preliminary blanching. If this is not possible, blanch in plenty of boiling water for 1–2 minutes and plunge at once into ice-cold water. Remove skins carefully. Cut the fruit in halves or slices and immediately cover with cold syrup. Leave a head space and put crumpled wax or greaseproof paper on top to keep them submerged. Seal and freeze.

To Use Thaw in the unopened container in the refrigerator or at room temperature and do not open until ready to serve or the fruit will begin to discolour. Slightly frosted is the best way of serving.

Raw for Dessert See page 138.

PEARS

These are not always satisfactory as they tend to be tasteless. Pears of good flavour can be stewed and frozen in the syrup. Honey or golden syrup can be used in place of sugar for extra flavour or preserved ginger cooked with the pears. Alternatively, freeze ones of good flavour as part of a fruit salad, putting them into the syrup immediately they are prepared. Otherwise treat them with ascorbic acid, see pages 135–6. See recipe page 160.

PINEAPPLE

This does not keep very well on its own but it is sometimes useful to freeze it for convenience. It is suitable for freezing as part of a fruit salad.

Dry Pack Peel, core and then slice, dice or grate according to the use to which it will be put. Sliced pineapple can be frozen dry without sugar and with pieces of Cellophane between each slice, then over-wrapped.

In Syrup Prepare as above in slices or dice and cover immediately with cold syrup.

To Use Either method, thaw in the unopened container and use while still frosty.

PLUMS (and Greengages).

Dry Pack Wash, grade for size and remove any damaged ones. They may be frozen whole for pies but it is better to halve or slice and remove the stones. Pack with or without sugar.

To Use For pies or tarts the fruit should be partially or fully thawed. Otherwise thaw, or cook from frozen, according to the way it is to be used. The plums can be used in place of fresh ones in any recipe.

In Syrup This gives a better flavour. Prepare as before and cover with cold syrup, leaving a head space. Seal and freeze.

To Use Thaw or cook from the frozen state.

Purée or Pulp See page 137.

Raw for Dessert See page 138.

RASPBERRIES

Dry Pack Pick over, removing hulls and any damaged berries. Avoid washing if possible. Unwashed berries can be frozen without sugar, otherwise mix with caster sugar before packing, or with small packs sprinkle the sugar on top.

To Use Thaw in the unopened container in the refrigerator or at room temperature. Use in place of fresh fruit in any recipe or serve as fresh fruit with cream. Serve while still frosted.

Purée and Pulps See page 137.

Juice See page 137.

TANGERINES see Citrus Fruit.

RHUBARB
For best results use young, tender stalks.

Dry Pack Wash and trim. The simplest method of freezing is to make a parcel of a bundle of even-lengthed stalks, using freezer wrap. Alternatively cut the rhubarb in pieces and freeze with or without sugar. Can also be blanched in syrup and frozen in syrup.

To Use. The sticks when partially thawed can be easily cut in pieces. For pies and tarts rhubarb should be partially thawed. Otherwise cook from the frozen state or thaw according to the method of use. It can be used in place of fresh rhubarb in any recipe.

Stewed Rhubarb Cook with sugar in the usual way, preferably in a casserole in the oven so that it keeps a good shape. Cool rapidly by standing the pan in cold water. Pack, leaving a head space, seal and freeze.

To Use To serve cold, thaw in the refrigerator or at room temperature; to serve hot, heat gently in a double boiler or in the oven.

Purée or Pulp See page 137.

STRAWBERRIES

Commercially frozen berries are usually whole but the flavour is often even better if they are frozen halved or sliced and for many uses this is preferable.

Dry Pack Wash if necessary and remove the hulls. To freeze small whole berries for garnishing purposes, put them on a tray, freeze uncovered and then pack in small bags or boxes. For serving whole with sugar, pack in boxes and sprinkle caster sugar over the fruit. Seal and freeze.

Alternatively cut the berries in half or slice them, using a stainless steel knife. Sprinkle them with caster sugar, mixing gently until the juice flows and dissolves the sugar. Pack, leaving a small head space, seal and freeze.

To Use Thaw whole berries in the container, slowly in the refrigerator and use while still frosty. Cut berries thaw in the unopened container either in the refrigerator or at room temperature. Use while still chilled.

In Syrup Use whole or sliced and freeze in cold syrup to cover. Thaw as above.

Purée and Pulp See page 137. Mash the berries with a fork, adding sugar to taste. Freeze in boxes. Thaw and use as it is or make it into a fine purée by sieving or blending.

Raw for Dessert See page 138.

Chapter Sixteen

ICES, COLD SWEETS AND PUDDINGS

ICES AND ICE-CREAMS

Many different types of commercial ice-cream are available and can be stored in the home freezer. If you use a lot of this type of ice-cream you may find it cheaper to buy a gallon or more at a time from the manufacturer. On the other hand, it may cost you just as much as buying in smaller and perhaps more convenient quantities.

Not all recipes for home-made ice-cream are satisfactory in the freezer, due to the faster freezing at a lower temperature. But if you have a recipe which is only satisfactory in the refrigerator it can still be stored in the freezer. With mixtures which are to be frozen direct in the freezer I find the proportion of cream is important. If there is too much the ice-cream has a grainy, fatty texture. Nor is it satisfactory if the ice is not allowed to thaw enough before serving. It needs to be softish and not still hard.

I have tested a number of different recipes which give satisfactory results in the freezer at 0°F (-18°C) and have included a selection of them here. None of them needs to be stirred during freezing, and all are easy and quick to make. I find a selection of ices is the best type of sweet course to keep in the freezer. The method of serving can be varied with different sauces, fruits, and other accompaniments to give variety.

Whether you buy commercial ices (and some genuinely prefer these) or make your own, a fairly quick turnover is recommended as an ice which will keep well for 1 month may change in texture with longer storage.

If you are going to keep commercial ices for more than 2–3 weeks the containers should be over-wrapped with freezer paper.

To Freeze Ices
They can be frozen in ice trays and stored in these provided the top is covered closely with a double layer of foil. Individual moulds are also suitable, small metal ones being the best. The ice can easily be turned out by dipping the mould in warm water for a moment to soften the outside slightly.

Larger moulds can be used to make elaborate ices with bands of different coloured and flavoured ices, or with ice-cream layered with fruit purée.

To Use
It takes a little experience to judge the thawing time required. This varies with the container and with the type of ice. I find it satisfactory to transfer a metal container from the freezer to the refrigerator (not the ice compartment) when the main course of a meal is about to be served. A 4–6 portion tray of Parfait is usually softish by the time it is required. In plastic containers I would give it longer, an hour or so in the refrigerator or $\frac{1}{2}$ hour at room temperature with the lid off the container. Whatever happens do not serve ices still frozen hard as they are then unpleasant to eat

and will not have a full flavour. If they are being served with a hot sauce, for example chocolate, they can naturally be harder to begin with.

Ices made with evaporated milk usually take a little longer to thaw than those made with cream and the ones containing eggs and cream thaw fastest of all.

BANANA ICE-CREAM

QUANTITIES for 8 or more

½ pt whipping cream (1 c or ¼ l): or 7 oz chilled evaporated milk (⅔ c or 2 dl)

Whip the cream or milk until thick and light but not stiff. Stiffly whipped cream spoils the texture of the ice-cream.

½ pt banana purée (1 c or ¼ l) or 4 medium bananas: 2 oz icing sugar (1 c or 60 g)

Make the purée by mashing very ripe bananas or putting them in the electric blender with the sugar. Fold the banana mixture into the whipped cream or milk and pour into freezer trays or rigid containers. Leave a little head space.

To Freeze
Cover freezing trays with a double lid of foil. Seal other containers and freeze without stirring.

To Use
Thaw in the ice compartment of the refrigerator or in the body of the refrigerator or at room temperature. It should be softish when served.

Serve with a chocolate sauce or any fruit sauce.

CHESTNUT ICE-CREAM

QUANTITIES for 8

1 lb canned chestnut purée (½ kg): 4 oz icing sugar (1 c or 120 g): 1 tsp vanilla essence: 1 Tbs rum

Combine these until smooth, preferably in an electric blender.

½ pt whipping cream (1 c or ¼ l) or use evaporated milk

Whip the cream or milk until light and combine it with the chestnut purée. Beat well.

To Freeze
Put in freezer trays or rigid containers, leaving a small head space. Cover the trays with a double lid of foil. Seal the containers. Freeze, without stirring.

To Use
Thaw in the refrigerator or in the ice compartment or at room temperature. Thaw until fairly soft. Serve with hot or cold chocolate sauce, see page 63.

ICED VANILLA SOUFFLÉ

QUANTITIES for 6

3 oz caster sugar (6 Tbs or 90 g): pinch of salt: ¼ pt water (½ c or 1½ dl)

Put in a small pan and boil rapidly without stirring for 5 minutes or until it reaches the soft ball stage (when a little dropped in water forms a soft ball).

3 egg whites

Beat the egg whites lightly and slowly pour in the syrup, beating all the time. Continue beating until the mixture cools.

¼ pt whipping cream (½ c or 1½ dl): Vanilla essence

Whip the cream lightly and fold it into the other mixture. Stir in the vanilla, about ½ tsp.

To Freeze
Pour into an ice tray or rigid container. Seal and freeze, without stirring. Alternatively freeze in individual waxed cups and serve in these.

To Use
Thaw in the refrigerator until softish.

FROZEN FRUIT SOUFFLÉ

Omit the vanilla and fold in ¼ pt fruit purée (1½ dl) before adding the whipped cream. To give a less rich ice, ¼ pt (½ c or 1½ dl) evaporated milk may be used in place of the cream.

Use a well flavoured fruit, for example, canned pineapple, loganberries, raspberries, strawberries or banana.

LEMON SHERBET OR SORBET

QUANTITIES for 4–6

3 *strips of yellow lemon rind: ¼ pt water (½ c or 1½ dl)*

Heat together until boiling.

3 *oz sugar (6 Tbs or 90 g)*

Strain the water on to the sugar, return to the pan and boil for 5 minutes.

2 *Tbs golden syrup or honey: 1 tsp gelatine soaked in a little cold water*

Add to the hot mixture and stir until all is dissolved.

¼ *pt lemon juice (½ c or 1½ dl)*

Add sufficient water to make the juice up to ½ pint (1 c or ¼ l). Add to the main mixture. Pour into the ice tray and freeze until it begins to set round the edges.

1 *egg white*

Beat until stiff and fold into the semi-frozen mixture. Freeze, without stirring.

To Store

Cover the trays with a lid of double foil.

To Use

Thaw in the refrigerator or freezing compartment until mushy. It can be kept for several days in the freezing compartment.

MOCHA ICE-CREAM

QUANTITIES for 6–8

2 *oz plain or bitter chocolate (60 g)*

Melt the chocolate by putting it in a basin over hot water.

½ oz sugar (1 Tbs or 15 g): 1 Tbs soluble coffee: ½ pt whipping cream (1 c or ¼ l): or 7 oz chilled evaporated milk (⅔ c or 2 dl)

Add the sugar and coffee to the chocolate together with 2 Tbs of the milk or cream. Stir and heat until smooth. Whip the remaining cream or milk until light but not stiff. Stir in the mocha mixture.

To Freeze
Pour into a freezing tray or rigid container leaving a small head space. Cover the tray with a lid of double foil. Seal the container and freeze, without stirring.

To Use
Thaw in the refrigerator or in the ice compartment or at room temperature, making sure it is a softish mixture when served. Serve plain or with whipped cream.

ORANGE PARFAIT

QUANTITIES for 4–8. This is very rich and sweet so serve just a small portion with a sauce or some fruit.

¼ pt orange juice (½ c or 1½ dl): 4 oz caster sugar (½ c or 120 g): pinch of salt

Heat these in the top of a double boiler, stirring until the sugar dissolves.

3 egg yolks

Beat in a small basin until the egg is thick and lemon-coloured. Add it to the orange mixture and beat constantly over hot water until the sauce thickens. Remove from the heat and stand the pan in a basin of cold water to cool. Whisk during cooling to make it light.

¼ pt whipping cream (½ c or 1½ dl)

Whip the cream until light but not stiff. Fold it into the cold orange mixture.

To Freeze

Put in small individual moulds or paper cups or put the whole mixture in a freezer tray or other freezer container. Seal and freeze, without stirring.

To Use

Thaw in the refrigerator until it is softish or thaw at room temperature.

MARSALA PARFAIT Substitute Marsala for the orange juice.

PINEAPPLE SHERBET OR SORBET

QUANTITIES for 8

8 *oz sugar* (1 *c or* 240 *g*): 1 *pt water or water and pineapple syrup mixed* ($\frac{1}{2}$ *l*)

Heat together until the sugar dissolves and then boil for 5 minutes.

1 *tsp gelatine*: 2 *Tbs water*

Soak together for a few minutes and then dissolve in the hot syrup. Cool.

2 *Tbs lemon juice*: 4 *Tbs orange juice*: $\frac{1}{2}$ *pt canned pineapple purée* ($\frac{1}{4}$ *l*)

Make the purée from crushed pineapple or pineapple pieces sieved or put in the blender. Add juices and pineapple to the cooled syrup. Pour into trays and freeze.

2 *egg whites*

Beat until stiff. When the ice is half frozen, fold in the egg whites and finish freezing.

To Store

Cover the trays with a lid of double foil.

To Use

Thaw to a mush in the ice compartment of the refrigerator where the mixture will keep in suitable condition for two or three days.

STRAWBERRY ICE-CREAM

QUANTITIES for 8 or more

½ *pt strawberry purée* (1 *c or* ¼ *l*)

To make the purée use either fresh ripe strawberries or thawed frozen ones or thawed frozen purée or pulp.

The quickest way of making the purée is to use an electric blender, otherwise rub the fruit through a sieve.

2 *oz icing sugar* (¼ *c or* 60 *g*)

Add to the fruit and stir until the sugar is dissolved. When a blender is used put the sugar and fruit in together.

¼ *pt double cream* (½ *c or* 1½ *dl*): ¼ *pt single cream* (½ *c or* 1½ *dl*)

Mix the creams and whip until light but not stiff. Fold into the purée combining gently but thoroughly.

To Freeze
Pour the mixture into freezing trays or rigid containers. Cover the trays with a lid of double foil and seal the containers. Freeze without stirring.

To Use
Thaw in the refrigerator or in the ice compartment or at room temperature. It should be softish when served.

Serve with some fresh or thawed whole or sliced berries and garnish with whipped cream.

ECONOMICAL ICE-CREAM USING DREAM TOPPING

QUANTITIES for 4

¼ *pt cold milk* (½ *c or* 1½ *dl*): 1 *Tbs icing sugar*: ½ *tsp vanilla essence*

Mix these in a basin.

1 *pkt Bird's Dream Topping* (2 *oz or* 56 *g*)

Add to the liquid and whisk until light and thick.

To Freeze
Pour into an ice tray or freezer box. Cover the tray with a lid of double foil or seal the box. Freeze without stirring.

To Use
Thaw in the refrigerator or at room temperature until softish. Use as a basis for sundaes with fruit, fruit sauces or chocolate sauce.

VARIATIONS

COFFEE RUM
Add 1 Tbs soluble coffee and 1 tsp rum with the sugar and milk.

FRUIT ICE
Make the basic recipe, omitting the vanilla. Before freezing fold in about 4 oz (120 g) of chopped canned or fresh fruit.

ICED LOLLIES
Special lolly moulds and sticks can be purchased for making these. Make them from canned or fresh fruit juices.

ST. IVEL ICE-CREAM MIX
This makes very good ice-cream in the freezer, though it is rather sweet for some palates. The makers recommend beating the ice-cream after it has been freezing for $\frac{1}{4}$ hour and I do this when using the freezer.

COLD SWEETS
These are very useful items to have in the freezer. It is very pleasant not to have to worry about making a sweet every day. A large quantity can be made at a time convenient to the cook and frozen for later use. Or, when you make some for one meal, double or treble the recipe and store the surplus. If the sweets are frozen in individual portions, then the family can have different sweets if they choose.

It is also very useful to make party sweets in advance and store them in the freezer.

There are a few types of cold sweets which are not as good when frozen as when merely refrigerated. Custards and milk puddings tend to separate. Recipes using much jelly lose their setting ability with long storage and clear jellies become cloudy. Those containing a small amount of gelatine freeze satisfactorily for short periods.

The recipes I have chosen for this section are representative of the type of sweet I have found to give good results over short storage periods of about one month. In fact most sweets can be stored satisfactorily in the freezer for these short periods.

APPLE SNOW
This recipe differs from many in that the snow is cooked after the egg white has been added. This gives a very smooth, stable mixture which freezes very well.

COOKING TIME: $\frac{3}{4}$–1 hr: QUANTITIES for 4–5

1 *lb cooking apples* ($\frac{1}{2}$ *kg*)

Wash the apples and bake them in a moderately hot oven E 400°F G 6 until they are soft. Rub through a sieve and keep the pulp hot.

4 *oz caster sugar* ($\frac{1}{2}$ *c or* 120 *g*): *juice of* 1 *lemon*

Add to the apple and stir until the sugar is dissolved.

2 *egg whites*

Use a quart size basin ($1\frac{1}{4}$ l) which will fit in the top of a saucepan. Beat the egg whites in the basin until they are stiff but not dry. Put the basin over boiling water, add the apple mixture and beat for 4–5 minutes. Remove from the heat.

To Freeze
Cool quickly by standing the pan in ice-cold water. Put the snow in one or two rigid containers, leaving a small head space. Seal and freeze.

To Use
Thaw in the refrigerator or at room temperature and then store in

the refrigerator. It may be spooned into individual serving dishes before it is quite thawed or serve in a glass bowl.

BAVAROIS OR BAVARIAN CREAM

COOKING TIME: about 10 mins: QUANTITIES for 6–8

4 egg yolks: $\frac{1}{2}$ pt milk (1 c or 2$\frac{1}{4}$ l) 2 oz sugar ($\frac{1}{2}$ c or 60 g): $\frac{1}{4}$ oz gelatine ($\frac{3}{4}$ Tbs or 7$\frac{1}{2}$ g)

Use the top of a double boiler or a basin which will fit in the top of a pan. Mix the ingredients together and cook over boiling water, stirring with a wooden spoon until the mixture begins to thicken and coats the back of the spoon.

Remove from the heat, stand the pan or basin in cold water to cool, stirring occasionally.

$\frac{1}{2}$ pt whipping cream (1 c or $\frac{1}{4}$ l): 1 oz caster sugar (2 Tbs or 30 g): vanilla or other flavouring (see below)

Whip the cream until it is thick and light but not stiff. Add the sugar and whisk lightly into the cold egg mixture. Flavour to taste.

To Freeze

Pour either into a fluted mould with a hole in the centre (border or ring mould) or into individual moulds. It can also be used to make a Charlotte Russe. In this case a mould is lined with sponge fingers, cutting them to make a flower shape at the bottom. Then fill the centre with the Bavarois mixture. Cover the tops of the moulds with Cellophane and over-wrap with double foil. Freeze.

To Use

Thaw overnight in the refrigerator or 2–3 hrs at room temperature. As this is a delicate mixture I find it better to unmould while it is still frozen by standing the mould in warm water for a minute or so. Leave the upturned mould as a cover until ready to serve the Bavarois.

It may be decorated with whipped, sweetened cream or with fruit.

Flavourings: Use one of the following:
Vanilla sugar or vanilla essence

Dissolve soluble coffee in the hot egg mixture.

Add 3 oz bitter chocolate to the egg mixture and allow to melt.

Add the finely grated rind of 1 orange or lemon to the egg mixture.

Flavour with almond essence, Kirsch, rum or any liqueur.

CHEESE CAKE

COOKING TIME: 45 mins: TEMPERATURE E 350°F G 4

QUANTITIES for a cake about 8 in (20 cm) diameter.

A foil pie plate is the best to use for this. Alternatively, line a heat resistant glass pie plate with foil, or line a deep sandwich tin with foil. Grease the foil lining thoroughly with butter or margarine.

1½ oz digestive biscuits (45 g)

Crumble the biscuits finely by putting them in the blender or by mincing them. Put about two thirds of them into the tin and tap and shake to coat the lining. Leave a thin loose layer on the bottom.

1 lb cottage cheese (2 c or ½ kg)

Put the cheese in the blender to make it smooth or rub it through a sieve.

2 oz melted butter (4 Tbs or 60 g)

Add to the cheese and mix well.

2 eggs

Beat thoroughly and stir into the cheese mixture.

2 Tbs cream: pinch of salt: 4 oz caster sugar (½ c or 120 g): grated rind of 1 lemon: 2 oz plain flour (6 Tbs or 60 g)

Add these gradually to the cheese mixture together with the juice of the lemon. Pour the mixture into the prepared tin. Cover with the rest of the crumbs.

Bake until the cake is set and brown.

To Freeze

If it has been baked in a foil-lined plate or tin, turn it out carefully on to a piece of foil. Allow to cool. When it is quite cold, double wrap in foil and freeze. Cover foil pie plate with foil lid.

To Use
Thaw in the refrigerator or at room temperature, still wrapped.
Remove from the wrapping and serve plain or with fruit and
cream.

COMPÔTE OF PEARS WITH HONEY AND GINGER

COOKING TIME: 15–30 mins: QUANTITIES for 6–8

*4 oz honey (4 Tbs or 120 g): 1 pt water (2 c or ½ l): 2 oz crystal-
lised ginger (60 g): 4 Tbs lemon juice*

Use a wide stew pan which will allow the pears to lie in a single
layer. Alternatively prepare and cook the pears in relays. Put the
honey and water in the pan and stir until the honey dissolves.
Add the sliced ginger and lemon juice.

2 lb dessert pears not quite ripe (1 kg)

Peel and halve the pears. Remove cores and immerse the fruit at
once in the syrup. They must be kept covered with syrup throughout cooking and freezing otherwise they will discolour. Poach
them gently until they are tender. Stand the pan in cold water to
cool them quickly.

To Freeze
Put in rigid containers with crumpled freezer paper on top of the
fruit to keep it submerged. Seal and freeze.

To Use
Thaw at room temperature and serve while the fruit is still chilled.
Serve plain or with cream.

DANISH RED PUDDING
To make and store when red currants and raspberries are in
season.

COOKING TIME: 15–20 mins: QUANTITIES for 12

*2 lb red currants (1 kg) and 1 lb raspberries (½ kg) or use 3 lb
fruit (1½ kg) mixed as desired: 1½ pt water (3 c or ¾ l)*

Wash the fruit. There is no need to remove stalks from the
currants. Put in a pan with the water and boil gently until all the

juice is extracted from the fruit. Strain through muslin or through a fine nylon or plastic sieve.

$1\frac{1}{2}$ *lb sugar (3 c or 750 g) or to taste*

Add to the juice and stir until it dissolves, heating gently if necessary.

To Freeze
Pour into a leak-proof container. Chill, seal and freeze.

To Use
Put in a pan and add 1–$1\frac{1}{2}$ pt water (2–3 c or $\frac{1}{2}$–$\frac{3}{4}$ l). Begin by adding the smaller amount of water. Heat gently and when it has thawed, taste to see if the flavour is strong enough to take more water.

3 oz potato flour or fecule (9 Tbs or 90 g): $\frac{1}{2}$ pt cold water (1 c or $\frac{1}{4}$ l)

Blend the potato flour with the cold water. Add some of the hot juice, mix and return to the pan. Stir until it thickens but do not boil. Pour into a serving dish or individual dishes.

Caster sugar: Blanched almonds: single cream

Sprinkle a fine layer of sugar on top to prevent a skin from forming as it cools. Cool and chill.

Decorate the top with almonds and serve the cream separately. The pudding should be the consistency of thin porridge.

FROZEN CHOCOLATE CAKE

QUANTITIES for 4

3 oz savoy fingers or dry sponge cake (90 g)

Use either an ice tray about 7–8 ins (18–20 cm) long or a 1 lb ($\frac{1}{2}$ kg) loaf tin of the shallow aluminium type. Cut a piece of non-stick lining paper to line the bottom and sides of the tin and give sufficient length to fold over the finished mixture.

Split the savoy fingers or cut the sponge cake into thin slices. Put a layer in the bottom of the prepared tray, with the rounded sides of the savoy fingers next to the paper. Keep the remaining pieces for the top.

2 eggs: 1 oz plain chocolate (30 g): 2 oz sugar (4 Tbs or 60 g): 2 Tbs milk

Separate the white and yolks of the eggs. Melt the chocolate over hot water or a gentle direct heat. Add the sugar, egg yolks, and milk and continue cooking over a gentle heat, stirring all the time, until the mixture thickens. Cool, stirring occasionally.

1 oz butter or margarine (2 Tbs or 30 g): 2 oz caster sugar (4 Tbs or 60 g)

Cream together until soft and light. Add to the chocolate mixture, combine thoroughly.

½ tsp vanilla essence or 1 tsp rum

Add the flavouring. Beat the egg whites until stiff and fold them into the other mixture. Pour it into the prepared tray and put another layer of cake on top. Fold over the spare paper to cover the top.

To Freeze
Cover with foil and freeze. Turn out of the container and over-wrap with foil. Return to the freezer.

To Use
Thaw for several hours in the ice compartment of the refrigerator and serve before it is completely thawed. Turn out on a dish and decorate with whipped cream. To serve, cut in slices.

FRUIT FOOL
Suitable for making with fresh, canned or frozen fruit or purée.

QUANTITIES for 4–6

½ pt thick fruit purèe (1 c or ¼ l): lemon or orange rind or spice to flavour: sugar, syrup or honey to sweeten

Sweeten and flavour the purée to taste and make sure it is cold before using.

If raw frozen fruit is used partially thaw it before making the purée, or before using a frozen purée. If the frozen fruit needs to be cooked to make the purée do this without thawing, only adding water if absolutely necessary. Use a gentle heat. Canned fruit should be drained thoroughly. To make the purée either sieve or blend the fruit.

¼ pt evaporated milk (½ c or 1½ dl) or ½ pt whipping cream (1 c or ¼ l)

Whip the milk until it is light and thick so that it stands up in peaks. Whip the cream until thick but not buttery. Fold in the fruit purée. Chill before serving, or freeze.

To Freeze
Spoon into rigid containers leaving a small head space. Seal and freeze.

To Use
Thaw several hours in the unopened container in the refrigerator or 1–1½ hrs at room temperature. Serve chilled.

FRUIT FOOL USING UNSWEETENED FROZEN PURÉE

QUANTITIES for 4

½ pt unsweetened fruit purée, partially thawed (1 c or ¼ l): 4–6 Tbs full-cream sweetened condensed milk: 2 Tbs lemon juice for flavouring or use spice or grated orange or lemon rind

Put the ingredients in the electric blender or whisk by hand until the mixture is smooth. Use the larger amount of milk if the fruit is sour or if a sweet fool is preferred. Very sour fruits like damsons may need additional sugar. Use icing or caster sugar.

Pour into individual dishes and chill before serving.

Whipped cream or whole fruit to garnish

Serve plain or garnished to taste.

ALTERNATIVE

Make a mousse by folding in 1 or 2 stiffly beaten egg whites.

FRUIT MOUSSE

QUANTITIES for 8–12

½ pt fruit purée (1 c or ¼ l): juice of 1 lemon: 1–2 oz caster sugar (2–4 Tbs or 30–60 g)

Use a fruit with a fairly strong flavour such as bananas, berries or prunes. Combine the fruit, sweetening and lemon thoroughly. If an electric blender is used put the whole fruit in with the other ingredients (1 lb fruit (½ kg) makes ½ pt purée).

½ pt whipping cream (1 c or ¼ dl) or use half double cream and half evaporated milk

Half whip the cream and fold it into the purée.

4 egg whites

Beat the whites until they are stiff enough to stand up in peaks. Fold into the other mixture.

To Freeze
Spoon into rigid containers or individual serving dishes. Leave a small head space. Seal and freeze.

To Use
Thaw unopened several hours in the refrigerator or 1½–2 hours at room temperature. Garnish with fruit or cream.

PEACHES IN WHITE WINE

COOKING TIME: 5–10 mins: QUANTITIES for 6

2 oz sugar (4 Tbs or 60 g): ½ pt sauterne (1 c or ¼ l): ¼ pt water (½ c or 1½ dl)

Combine these in a large shallow pan, stirring until the sugar dissolves.

6 medium ripe yellow peaches

Skin the peaches carefully. Cut in half and remove stones. Place them in the syrup as they are prepared. Cover the pan and cook over a low heat until the peaches are just tender.

To Freeze
Lift them out carefully with a spoon and put them in a rigid container. Pour over the syrup making sure there is enough to cover them completely. Put a piece of crumpled freezer paper on top of the fruit to keep it submerged. Seal and freeze.

To Use
Thaw in the unopened container in the refrigerator or at room temperature. To hasten thawing, the container may be stood in a bowl of warm water. Serve while the fruit is still chilled.

  3 *Tbs rum or Maraschino*

Just before serving pour this over the peaches.

STEAMED PUDDINGS
Both suet puddings and sponge puddings are suitable for freezing. Make and cook in the usual way, using foil pudding basins if possible. Cool, wrap and freeze.

To Use
Thaw at room temperature for about 6 hours and then heat by steaming or boiling for 45 minutes to 1 hour or until heated through.

   Sponge puddings can be heated in the oven at E 350°F G. 4 for 15–20 minutes for small ones.

FRUIT CRUMBLES
Make as usual and freeze raw. Or make the crumble topping and freeze it separately to use with frozen fruit. Assemble while fruit and topping are still frozen.

To Use either method.
Bake at E 375°F G 5 for about 1 hour.

BAKED SPONGE PUDDINGS
These can be treated in the same way as steamed sponge puddings, see above.

   Others can be treated as cakes, see page 183.

   Upside-down cakes are particularly successful in the freezer and the following recipe is a typical example.

PINEAPPLE UPSIDE-DOWN CAKE OR PUDDING
  COOKING TIME: 35–45 mins: TEMPERATURE: E 375°F G 5

  QUANTITIES for a tin about 7 × 9 ins and 1½ ins deep, or a similar sized square or round tin (20 × 24 × 3 cm)

1 *oz butter* (3 Tbs or 30 g): *fine light brown sugar: canned pine-apple slices* (1 lb tin or ½ kg): *glacé cherries*

Melt the butter in the cake tin and sprinkle thickly with sugar. Drain the pineapple and arrange the rings on top of the sugar cutting some to fill the corners. Put a cherry in the middle of each complete ring.

3 *oz butter or margarine* (6 Tbs or 90 g): 3 *oz caster sugar* (6 Tbs or 90 g): 1 *egg*

Cream the fat and sugar together and beat in the egg.

6 *oz self-raising flour* (1¼ c or 180 g): *pinch of salt.*

Sift these into the creamed mixture and stir to blend. Spread the mixture gently over the fruit. Bake until a skewer inserted in the middle comes out clean. Turn the cake upside-down on a tray and leave to become quite cold.

To Freeze
It can be frozen on the tray and then wrapped, or cut in portions and pack in a rigid container with Cellophane between pieces. Seal and freeze.

To Use
Thaw at room temperature in the unopened container. Serve as a cake or pudding with cream handed separately.

ALTERNATIVE
Use a packet sponge mix instead of the above recipe. A sponge sandwich mixture is about the right size.

Chapter Seventeen

PASTRY AND PIES

Commercially frozen pastry makes a useful item for storing. The puff pastry saves much time, for making it at home is a time-consuming job. On the other hand the fats used in commercial pastry are usually tasteless and, if you are a good pastry maker, you will get better results making your own puff pastry using butter or good quality margarine.

Most pastry is satisfactory frozen raw or cooked, though hot water pastry for raised pies should be used only with cooked pies.

TO FREEZE UNROLLED PASTRY
Make in the usual way. Shape it into a flat oblong which will freeze and thaw faster than a ball or lump of pastry. Wrap it closely in freezer paper and freeze.

To Use
Thaw it at room temperature until it is soft enough to roll.

TO FREEZE RAW AND COOKED PIES AND TARTS
Basic directions for doing this are included in the following pages. See also Chicken Pie page 114.

CHOUX PASTRY
Make in the usual way.

To Freeze Raw
Shape on a tray, freeze and pack.

To Use
Put on a baking tray, thaw at room temperature 15–30 minutes depending on the size. Bake as usual.

To Freeze Cooked
They may be filled with whipped cream or ice-cream. Unfilled ones pack in bags or boxes. Freeze filled ones on trays before packing.

To Use
Remove filled ones from the freezer and thaw them at room temperature. Fill others while frozen and the filling will help to thaw them.

To Freeze Small Savoury Puffs
Freeze unfilled.

To Use
Bake at E 250°F G ½ for 10–15 mins or until crisp. Cool and fill with the savoury mixture.

FRUIT FLANS
Unfilled cases can be frozen raw or cooked. Cooked ones need some protection against damage. It is advisable to freeze them un-wrapped and then stack in a box.

Raw ones should also be frozen unwrapped and can be stacked for storage. Put Cellophane or polythene paper between each.

Fill with fresh, thawed frozen or canned fruit in the usual way.

Baking
Raw Cases Bake at E 400°F G 6 for 20–25 mins.
Cooked cases Thaw at room temperature for about 1 hr.

QUICHES AND SAVOURY FLANS
Make in the usual way. Cool, chill and freeze.

To Serve Cold
Thaw in the refrigerator for at least 6–8 hrs.

To Serve Hot
Heat in a moderate oven E 350°F G 4 for 20 mins or until well heated through.

FRUIT PIES
Use foil pie dishes for these and make the pie in the usual way.

When using apples, which tend to discolour if the pie is frozen raw, blanch the sliced apples for 2 mins in boiling water and then chill in cold water. Drain. Then put in the pie dish in the usual way with sugar to taste. It is better not to cut the apples too finely, average-sized ones in about 12 slices.

Whether the pie is being cooked first or frozen raw, freeze un-wrapped. When frozen, wrap and store, but avoid putting heavy packages on top of the crust.

There is not much point in cooking the pie first, unless it is to be served cold.

Cooked Pies
To serve cold thaw overnight in the freezer or 5–6 hrs at room temperature.

To serve hot, heat at E 375°F G 5 for 30–40 mins.

Raw Pies
Cook at E 400°F G 6 for 1 hr or until the pastry is lightly coloured and the fruit cooked. Cut a vent in the top of the crust after the pie has been in the oven for about 15 mins.

MEAT PIES (single crust)
Meat pies such as steak and kidney or veal and ham can be frozen successfully. Below are details of the three ways in which this may be done, the choice of method depending on personal preference and to some extent on the size of the freezer, for completed pies are bulky to store.

1 *Freezing Baked Pies*
The pie is made and baked in the usual way and should be topped up with stock if necessary as you would for a pie to be served cold. Cool the pie as quickly as possible, wrap and freeze.

To Serve cold
Thaw in the refrigerator or at room temperature and then store in the refrigerator until required. Thawing takes several hours, even at room temperature.

To Serve Hot
Bake the pie in a fairly hot oven for about 1 hour.

2 *Freezing Raw Pies*
The pie is made in the usual way except that a vent is not cut in the top, nor is it brushed with glazing liquid. Wrap and freeze.

To Cook while Frozen
Flaky or Puff Pastry Bake at E 475°F G 9 for 25 mins then E 375°F

G 5 for 2 hrs or whatever the normal cooking time is with an unfrozen pie.

Before cooking brush the pastry with an egg wash or milk and cut a vent in the top when the pie has been in the oven for about 25 mins.

Short Pastry Bake at E 425°F G 7 for 15 mins then E 375°F G 5 for 2 hrs or whatever the normal cooking time with an unfrozen pie. Before cooking brush the pastry with an egg wash or milk and cut a vent in the crust after the pie has been cooking for about 15 mins.

In either case if the pastry shows signs of becoming too brown, cover it with a piece of foil.

Pie-dishes

Oblong ones waste least space in the freezer. Use the special foil dishes sold for this purpose.

3 *Freezing the Ingredients Separately*

The chief merit of this method is that it takes less freezer space, and special foil dishes are not required. Prepare the ingredients for the filling but do not add any liquid. Wrap in freezer paper making a fairly flat oblong parcel for quick freezing and thawing.

Make the pastry and freeze it in a flat block. Thaw the meat and pastry at room temperature. Make up the pie in the usual way.

FREEZING RAISED MEAT PIES

Make and bake in the usual way. Cool as quickly as possible. Then chill in the refrigerator. Wrap closely and freeze.

To Use

Thaw in the refrigerator or at room temperature and then store in the refrigerator until required.

FREEZING PASTIES AND TURNOVERS

Make in the usual way and freeze either raw or cooked. Chill cooked ones thoroughly and rapidly before wrapping and freezing.

To prevent damage to the crust, freeze raw ones on a tray before wrapping them.

To Use

Brush raw ones with egg wash and bake at E 400°F G 6, for 50–55 mins.

Thaw baked ones for 12 hours in the refrigerator for serving cold.

To serve baked ones hot, re-heat for 20 minutes in a fairly hot oven.

PLATE PIES, FRUIT TARTS OR DOUBLE CRUST PIES

No special recipes are needed for these but certain alterations in the method of preparation can improve results.

Sometimes the bottom crust in frozen fruit pies is inclined to be soggy. Various ways of preventing this have been recommended, such as dusting the inside of the lower crust with flour, or brushing it with egg white or melted fat. I find it even better to use a pastry containing egg, see recipes page 173.

Pies can be baked before freezing or frozen raw. There is not much point in baking first unless the pie is to be served cold. However, with fruit that discolours easily, for example apples, baking before freezing makes it unnecessary to pre-treat the apples to prevent browning during storage.

Pie Plates

Foil pie plates are sold in a range of sizes and are very convenient to use. They can usually be washed and used again.

Line oven glass pie plates with foil before making the pie, freeze the pie unwrapped, remove from the plate, wrap and return to the freezer. When the pie is to be served return it to the plate for cooking and serving. The foil lining can be peeled off before putting the pie in the plate for the second time.

Preparation

This is normal except that a vent is not cut in the top crust until the pie is to be cooked, easiest to do when the crust has thawed (about 15–20 mins after it has been put in the oven).

Pies intended for freezing uncooked and made with fruit such as apples or apricots should have the fruit treated to prevent discoloration during storage.

This can be done by blanching the sliced apples in boiling water for 2 minutes, then cooling in cold water. Drain and dry before putting in the pie.

Alternatively, toss the apples in a solution of ascorbic acid. Dissolve 500 milligrams of ascorbic acid tablets in $\frac{1}{4}$ pt cold water ($\frac{1}{2}$ c or $1\frac{1}{2}$ dl). This quantity is sufficient for 1 lb of fruit.

Using Frozen Fruit
There is no reason why you shouldn't make a pie with frozen fruit and then re-freeze if it is a convenient thing to do.

Perhaps more sensible is to make the pie when it is required using frozen fruit and frozen or freshly made pastry. Frozen fruit needs to be thawed only enough to separate the pieces. Then make the pie in the usual way.

To Freeze
The crust of baked pies needs some protection from damage in the freezer. It is a good plan to cover the crust with a foil pie plate, then over-wrap the whole or put it in a polythene bag.

Raw pies should be frozen unwrapped to avoid damage to the crust, then wrap and return to the freezer. Even when they are frozen it is advisable to avoid putting very heavy foods on top of the crust.

To Use
Cooked pies to be served cold, thaw at room temperature for 5–6 hours.

To serve hot, heat in the oven at E 375°F G 5 for 30–40 minutes.
Raw pies Cook at E 400°F G 6 for 20 mins. Then 350°F G 4 for 35 mins or until the fruit is tender and the crust lightly browned.

SHORT PASTRY WITH EGG
Ordinary short pastry is very satisfactory for freezing but this one with egg added is even better, especially for tarts made with juicy fruits and moist fillings.

QUANTITIES for a 7–8 in (20 cm) double crust tart or two flans.

8 *oz plain flour* (1$\frac{1}{2}$ *c or* 240 *g*): *pinch of salt*

Sift these into a mixing bowl.

2 *oz butter or margarine* (4 *Tbs or* 60 *g*): 2 *oz lard or cooking fat* (4 *Tbs or* 60 *g*)

172

Cut the fat in small pieces and rub it into the flour until the mixture looks like fine breadcrumbs.

1 *egg*

Beat a little and then mix the egg into the flour. Use the hands to work in dry flour, adding water only if absolutely necessary. Work until smooth and pliable.

To Freeze
It can be frozen in a piece for later use, or roll and make up the pie or tart and then freeze.

SWEET PASTRY WITH EGG

QUANTITIES for one 7–8 in double crust pie or two 7–8 in flans

8 *oz self-raising flour* (1½ *c or* 240 *g*): *pinch of salt:* 3 *oz butter or margarine* (6 *Tbs or* 90 *g*)

Mix the flour and salt and rub in the fat.

2 *oz caster sugar* (4 *Tbs or* 60 *g*): 1 *egg*

Add the sugar. Beat the egg and use it to mix the dry ingredients, working and kneading with the hands until it makes a smooth mixture. Do not add water unless absolutely necessary, for example if a very small egg has been used. Roll out to between ⅛ to ¼ inch and use in the normal way, for flans, sweet tarts and pies and small tarts.

To Freeze
Roll and make the pie or tart. Then freeze.

CONTINENTAL APPLE FLAN

COOKING TIME: ¾–1 hr: QUANTITIES for a 7½ in (18 cm) flan

TEMPERATURE: E 400°F G 6

8 *oz cooking apples* (240 *g*): ¼ *tsp grated lemon rind:* ½ *oz butter* (1 *Tbs or* 15 *g*): 1 *oz sugar* (2 *Tbs or* 30 *g*)

Peel, core and slice the apples. Mix them with the other ingredients and stew gently until pulpy. Add water only if necessary to prevent burning. Leave to cool.

Short pastry using 4 oz flour or use 6–8 oz ready-made or ready-mix

Roll this to line a 7½ in (18 cm) foil pie plate. Spread the cold apple pulp over the bottom of the pastry.

8 oz cooking apples (240 g): gran sugar

Peel, core and slice the apples very thinly. Arrange them on the apple pulp in an overlapping spiral design, beginning from the centre. Sprinkle with sugar.

Bake until the pastry is firm, about ½–¾ hr.

2 oz apricot jam (¼ c or 60 g): 1 Tbs water

Heat together in a small pan until runny. As soon as the flan comes out of the oven, brush the jam over the fruit. Leave to become cold.

To Freeze
Freeze unwrapped and then cover with double foil or put in a polythene bag, protecting the top with a piece of Cellophane. Return to the freezer.

To Use
Thaw at room temperature for 3–4 hrs and serve cold. Thawing time will depend on the temperature of the room. It does not hurt the flan to thaw some time before it will be used, so begin thawing in plenty of time.

MINT AND CURRANT PATTIES
An old English recipe, unusual and delicious.

COOKING TIME: 25 mins: TEMPERATURE: E 400°F G 6

QUANTITIES for 12 patties

8–12 oz pastry, short, flaky or puff using 6–8 oz flour (180–240 g)

Roll the pastry as thinly as possible and cut in rounds to line shallow patty tins. Cut an equal number of rounds for the lids.

½ oz fresh mint leaves (¼ c chopped or 15 g): 2 oz brown sugar (¼ c or 60 g): 2 oz currants (¼ c or 60 g)

Wash and dry the leaves before chopping them finely. Add them to the fruit and sugar and mash into a paste. Alternatively put the unchopped leaves, sugar and currants in the blender. Put a little into each patty case, moisten the edges with water, press on the lids.

To Freeze
Cover with a single layer of foil and freeze. Remove from the tins and pack in a box or bag, seal and return to the freezer.

To Use
Return to the patty tins. Brush with milk or water and sprinkle with sugar. Cut a small slit in the top of each. Bake while still frozen. Serve warm or cold.

MINCE PIES
Make these well in advance of Christmas and bake a few every time you want to serve some. They should keep in good condition for 2 months.

COOKING TIME: 20–30 mins : QUANTITIES for 2–2½ doz pies

TEMPERATURE: E 425°F G 7

2 lb short pastry (1 kg): 2 lb mincemeat (1 kg)

Roll the pastry to about ⅛ inch thick (3 mm). Use round cutters. Choose one the size of the tops of bun tins. Use this to cut the lids and a slightly larger one to cut the bottoms. Press the larger pieces into the patty tins. The pastry edge should come level with the top. Fill with mincemeat.
   Brush the edge of the pastry with water and press on the tops.

To Freeze
Freeze in the tins with a piece of foil over the top. Remove from the tins and pack the pies in boxes or parcels with Cellophane between layers. Seal and store. Do not put any heavy packages on top of them.

To Use
Take out the number required and re-seal the parcel. Return the pies to the original tins and cut a small slit in the top of each.

Brush with water and sprinkle with caster sugar. Bake until lightly browned. Remove from the tins on to a cake rack. Serve while still warm.

SMALL SAVOURY PIES
These can be made in the same way as mince pies using a cooked savoury filling of minced or chopped meat, poultry or game in a good sauce, or use flaked fish in a Béchamel sauce.

Chapter Eighteen

BREAD, CAKES AND BISCUITS

Most of these will freeze satisfactorily and most can also be kept well by other means. Whether it is worth while using freezer space for them depends on individual needs. If you do your own bread making and cake baking, then to make more than will keep satisfactorily by other methods and to freeze the surplus is a time saver. Or if yours is a small family with problems of preventing staling of bread and cakes before they can be used up, then freezing these items can be an economy as well as a convenience.

Bread and Yeast Goods

UNBAKED DOUGHS
These can be frozen satisfactorily and stored for up to 2 weeks.

The dough is prepared in the usual way and given its first rising. Then it is shaped, wrapped and frozen. It helps to prevent a crust from forming if the dough is brushed with oil or melted fat before freezing.

Separate layers of rolls and buns with freezer paper.

To Thaw
Do this as quickly as possible in a warm place. The second rising takes place during thawing. Bake in the usual way.

BAKED BREAD AND ROLLS

Most people find this is a more convenient method than freezing the uncooked dough. They keep longer too, up to 1 month. This way it is possible always to have fresh bread simply by thawing or warming in the oven.

Shop-made bread and other yeast goods are very satisfactory provided they are freshly baked but quite cold.

Always cool all products thoroughly before freezing them otherwise they will go soggy when thawed.

To Freeze

Bread does not stale in the freezer provided it is properly wrapped. If it is to be stored for only about a week the waxed wrapper in which sliced bread is sold will give sufficient protection provided the seal has not been broken. If you intend to take out only a few slices at a time it is better to over-wrap with a polythene bag and tie. You can then re-wrap to a smaller size, excluding air.

Long thin loaves usually store more satisfactorily than short fat ones. All kinds of buns, Danish pastries and similar goods are very satisfactory.

To Use

Unless the bread is going to be heated or toasted, thaw it at room temperature in the wrapper. If removed from the wrapper it tends to become soggy with condensation. Thawing a 1 lb loaf will take 7 hours while a 2 pound loaf can take up to 12 hours. Thawing can be speeded up by putting the loaf, foil wrapped, in a fairly hot oven, E 375°F G 5 for up to 30 minutes. It should then taste like freshly baked bread.

If the loaf is sliced thawing can be speeded up by removing the number of slices required, prising them apart and spreading them out on a rack.

Sliced bread can be toasted while still frozen and takes very little longer than fresh bread.

Yeast rolls are better if heated to thaw them, preferably foil wrapped to prevent drying of the outside. Heat at E 350°F G 4 for 10–15 minutes. Alternatively, take a bag of rolls out of the freezer at night and they will be thawed to give fresh rolls for breakfast. This method is particularly successful with soft rolls.

With crusty rolls the crust tends to soften and heating is a better method for retaining the crisp crust.

Buns and yeast pastries can be thawed either way and are generally very satisfactory thawed in the unopened bag or box.

BREADCRUMBS

Breadcrumbs tend to develop mould if stored at room temperature or even in the refrigerator. To store them in the freezer is very satisfactory and a great convenience. So often one wants breadcrumbs for a recipe and there is no suitable stale bread available. Packets of dried crumbs are not always a satisfactory alternative.

Breadcrumbs can be frozen in small bags and used while still frozen. Remove the amount required and re-seal the bag.

Buttered Crumbs ready for topping baked savoury dishes are a good store to have in the freezer. It saves much time to make a good supply at once and store the surplus. They can be used frozen and will thaw quickly on top of the food.

> 1 *pt fresh white breadcrumbs (2 c or ½ l): 1 oz butter or margarine (2 Tbs or 30 g)*

Melt the butter or margarine and add the crumbs. Stir until they are well coated, dry and separate. Cool.

To Freeze

Put in a polythene bag, seal and freeze.

To Use

Take out as much as required and re-seal the bag. They may be used while still frozen for gratin dishes and topping other savoury and sweet dishes. They give a better appearance and flavour than plain breadcrumbs.

YEAST

Both fresh and dried yeast keep longer in the freezer than by any other means but do not store fresh yeast for very long periods (many months) or it may lose quality.

BAKING POWDER BREAD AND SCONES

These are all satisfactory for freezer storage, better if a fairly moist mixture is used. Avoid over-baking.

To Freeze

Cool as quickly as possible on a rack. Wrap loaves tightly with freezer paper or put in polythene bags.

Freeze small goods on a tray unwrapped, and then pack them closely in bags or rigid containers. Separate layers with Cellophane paper.

To Use

Thaw loaves in the wrapper at room temperature for 6–7 hours. If the bread has been sliced before freezing take out the number of slices required and thaw them on a rack.

Thaw scones and other small goods in the oven at E 375°F G 5 until heated through. They will keep more moist if covered with foil during heating. Use while still warm.

CRANBERRY LOAF

COOKING TIME: 1 hr: TEMPERATURE: E 350°F G 4

QUANTITIES for a 1 lb ($\frac{1}{2}$ kg) loaf tin

4 oz cranberries (1 c or 120 g): $\frac{1}{2}$ Tbs grated orange or lemon: 4 Tbs orange juice: 1 oz chopped nuts ($\frac{1}{4}$ c or 30 g)

Line the loaf tin with non-stick paper or foil. Chop the cranberries coarsely. Prepare the rind and juice and chop the nuts.

8 oz self-raising flour (1$\frac{1}{2}$ c or 240 g): $\frac{1}{4}$ tsp salt: 1 oz butter or margarine (2 Tbs or 30 g)

Sift the flour and salt into a basin and rub in the fat.

4 oz sugar ($\frac{1}{2}$ c or 120 g)

Add to the flour mixture.

1 egg (large)

Beat the egg well and add it with the orange juice and the rind. Mix until smooth. Add the nuts and cranberries and put the mixture in the prepared tin, smoothing the top.

179

Bake until lightly browned and firm. Turn out on to a cake rack and cool.

To Freeze
Leave overnight or several hours before slicing. Pack into a polythene bag and freeze.

To Use
Remove the number of slices required and thaw them at room temperature. Serve spread with butter or margarine.

FRUIT LOAF

COOKING TIME: 2 hrs: TEMPERATURE: E 350°F G 4

QUANTITIES for a 2 lb (1 kg) loaf tin or 6 in (15 cm) cake tin

Grease the tin and put a piece of non-stick paper at the bottom or line the tin with foil.

12 *oz self-raising flour* (2¼ *c or* 360 *g*): 1 *tsp mixed spice*: ¼ *tsp salt*

Sift these into a mixing bowl.

4 *oz butter or margarine* (½ *c or* 120 *g*)

Rub into the flour.

4 *oz sugar* (½ *c or* 120 *g*): 1 *oz chopped peel* (3 *Tbs or* 30 *g*): 8 *oz mixed dried fruit—currants, sultanas, raisins* (1½ *c or* 240 *g*): 1 *oz chopped almonds or walnuts* (¼ *c or* 30 *g*)

Add to the flour and mix.

1 *egg*: ¼ *pt milk* (½ *c or* 1½ *dl*): 3 *Tbs orange or lime marmalade*

Beat the egg and add it to the flour mixture together with the milk and marmalade. Mix very thoroughly. Put into the prepared tin and smooth the top. Bake until a skewer inserted in the middle comes out clean.

To Freeze
Turn out of the tin and leave to become quite cold. If the loaf is to be sliced before freezing, leave it 24 hours before slicing, otherwise it tends to crumble. Wrap and freeze.

To Use
Thaw in the wrapper or remove the number of slices required and thaw on a rack. Spread with butter or margarine.

TREACLE LOAF

COOKING TIME: 1 hr: TEMPERATURE: E 350°F G 4

QUANTITIES for a 1 lb (½ kg) loaf

8 *oz self-raising flour* (1½ *c or 240 g): ½ tsp salt: ½ tsp ground ginger: ½ tsp mixed spice*

Grease the tin and line the bottom with non-stick paper. Sift the dry ingredients into a basin.

2 *oz margarine* (4 *Tbs or 60 g): 4 Tbs milk*

Warm these together just to melt the margarine. Cool slightly.

2 *Tbs treacle: 2 oz fine brown sugar* (4 *Tbs or 60 g): 1 egg (large)*

Beat the egg and add to the margarine with the treacle and sugar. Add to the dry ingredients and mix well. Pour into the prepared tin. Bake until firm. Turn out on a cake rack to cool.

To Freeze
When quite cold, slice and pack in a polythene bag. Seal and freeze.

To Use
Remove the number of slices required and thaw at room temperature. Serve spread with butter or margarine.

SCONES
Make in the usual way. Allow to become quite cold.

To Freeze
Freeze on trays unwrapped.
    Pack close in polythene bags without squashing but excluding as much air as possible. Seal and freeze.

To Use
Put the scones on a baking tray and cover with a lid of foil. Bake

for 10–15 mins at E 400°F G 6 or until well heated. Serve hot or warm.

Alternative method: split the scones and heat them under the grill. They can be toasted from the frozen state.

DROP SCONES, SCOTCH PANCAKES OR PIKELETS
Make these in the usual way but cool them on a rack instead of in a cloth.

To Freeze
Freeze on trays unwrapped. Stack them one on top of the other and put in polythene bags. Seal and freeze.

To Use
Spread them out on a tray and put in a cold oven set at E 400°F G 6 for 10–15 mins or until warmed through. Serve while still warm.

VARIATION
Make pancake-size drop scones and, when they are hot, pile them up with butter between layers and serve with syrup or honey sauce in place of a pudding. Cut the pile in wedges to serve.

PANCAKES
Make in the usual way and cool on a rack. Pack with Cellophane between each. Place in a polythene bag or parcel of freezer wrap. Seal and freeze. They may be filled before freezing.

To Use
Separate frozen unfilled pancakes and spread them out on trays. Cover with foil and bake at E 400°F G 6 for 10–15 mins. Serve with sugar and lemon or make savoury filled ones. Heat filling separately. This could be a frozen mixture, for example Chicken à la King or Creamed Chicken.

To Use Filled Pancakes
Heat as above but allow a little longer for the filling to thaw and heat. Test one before serving.

DOUGHNUTS
Prepare in the usual way and fry.

To Freeze
Freeze on a tray, uncovered, then pack, seal and freeze. Put Cellophane paper between the layers.

To Use
Heat at E 400°F G 6 for about 10 mins or until hot.

CAKES
Most cakes can be frozen satisfactorily. Spice cakes, however, should be stored for about two weeks only or there will be flavour changes.

The majority of cakes keep equally well in an airtight container or wrapper and there is not much point in freezing them.

Those which are well worth freezing are the light sponge cakes which tend to stale fairly quickly, and cakes with cream fillings and other moist, easily perishable fillings.

Though cakes can be frozen uncooked it is usually better to bake them first. Avoid over-baking.

To Freeze
Cool as quickly as possible making sure the cake is cold in the centre. If you will need to use only a little of the cake at a time it is more practical to cut it in portions. Once it is thawed frozen cake stales more quickly than fresh.

Delicate cakes and iced cakes are better if frozen unwrapped on a tray. Then cover with freezer wrap. The same applies to small cakes which might easily be damaged by close packing. Separate layers and slices with Cellophane paper.

To Use
Thaw at room temperature. Large cakes will take up to 12 hours, small cakes will take 2–3 hours, less time in a warm place.

Faster thawing can be done in the oven, the cake foil-wrapped, at E 375°F G 5 for 10–15 mins but the cake will stale very quickly on cooling.

CAKE CRUMBS Store in the same way as breadcrumbs, see page 178.

ICINGS

Butter icings, butter creams and glacé icings are the most satisfactory, see recipes pages 188-9.

Boiled icings and egg white icings tend to become dry and crumbly if stored for any length of time though they are satisfactory for short storage periods.

Iced cakes should be frozen unwrapped and then wrapped and returned to the freezer.

To Use

Loosen the wrapper so that it does not touch the icing, or remove it altogether. The icing may appear moist at first but will become dry when the cake is thawed.

FILLINGS

Whipped cream and ice-cream make satisfactory fillings for sponges.

Jam is better put in after thawing. Swiss rolls can be rolled up with greaseproof or Cellophane paper inside and then filled with jam when they are thawed.

Butter creams make suitable fillings. Fresh fruit fillings tend to make a cake soggy but dried fruit fillings are satisfactory.

CHOCOLATE SWISS ROLL

This is a little different from the conventional Swiss roll. It is excellent in the freezer, especially when filled with butter cream flavoured with vanilla or rum.

COOKING TIME: about 10 mins: TEMPERATURE: E 400°F G 6

QUANTITIES for 1 Swiss roll

Grease the tin and dust it with flour.

   3 eggs: 2 oz fine brown sugar ($\frac{1}{4}$ c or 60 g)

If the eggs have come straight out of the refrigerator or cold larder warm them by standing the mixing bowl in hot water.

Beat the eggs until light and then beat in the sugar. Continue to beat until the mixture is so thick that when the beater is withdrawn the mixture takes several seconds for the impression to go.

1 oz cornflour (3 Tbs or 30 g): 1 oz cocoa powder (3 Tbs or 30 g): pinch of salt: ¼ tsp vanilla essence

Sift the cornflour, cocoa and salt into the egg mixture and fold in thoroughly. Add the vanilla. Pour into the prepared tin and spread evenly. Bake at the top of the oven until it feels springy when pressed in the centre.

Turn the roll on to a piece of greaseproof paper which has been dusted with caster sugar. Trim off the edges and roll up with the paper inside. Put on a rack to become cold.

Unroll and fill with whipped cream or butter cream flavoured with vanilla or rum, see page 187.

To Freeze
Freeze the roll unwrapped and then wrap closely in foil or freezer paper.

To Use
Thaw at room temperature in the original wrapper.

CHOCOLATE VENETIANS

Chocolate Swiss roll recipe above: vanilla flavoured butter cream, see page 187; granulated sugar.

Make the Swiss roll but cool it flat on a cake rack. When cold cut it into three even strips lengthwise. Sandwich these together with the butter cream and ice the top thinly with the same cream. Sprinkle with granulated sugar.

Cut in about 10 portions.

To Freeze
Put on a tray and freeze unwrapped. Then pack in a box with Cellophane paper between the layers or pack in a single layer.

To Use
If in a single layer thaw unopened at room temperature. Otherwise separate the layers and thaw at room temperature.

GENOESE SPONGE OR PASTRY

QUANTITIES for two 7½ in (18 cm) sandwich tins, or one 9 in ×

$1\frac{1}{4}$ in deep (22 × 3 cm) tin, or an oblong or square tin 10 in × 8 in (25 × 20 cm) or 9 in square (22 cm)

COOKING TIME: 20–30 mins: TEMPERATURE: E 350°F G 4

Grease and flour the tin and line the bottom with a piece of non-stick lining paper.

3 eggs: 3 oz caster sugar (6 Tbs or 90 g)

If the eggs are cold, warm the mixing bowl. Beat the eggs until they are light. Add the sugar and continue beating until the mixture is so thick that when the beater is removed the impression takes some time to level out again.

3 oz plain flour ($\frac{2}{3}$ c or 90 g): 2 oz melted butter (60 g)

Fold the flour into the egg mixture, adding the butter, warm but not hot, at the end. Make sure it is thoroughly blended in.

Pour the mixture into the prepared tins and bake until the centre feels springy when pressed. Allow to cool in the tins until it begins to shrink from the sides. Turn out carefully and remove the paper.

When cold it can be cut in layers to make a layer cake or used for small fancy cakes. The mixture is also suitable for a sponge flan, ice box cakes and many other purposes.

To Freeze

It can be frozen plain and then thawed before being iced and decorated or finish the cake before freezing, see page 183.

To Use

For thawing iced cakes see page 184.

For the plain sponge thaw in the wrapping and then fill with whipped cream, jam, lemon curd or any other filling. Alternatively cover a single sponge with whipped cream mixed with fresh raw fruit. Another way is to arrange fruit on top as with a flan and then glaze the fruit and decorate with whipped cream.

Sponge flan, thaw at room temperature for 2–3 hrs.

LEMON GÂTEAU

Genoese sponge baked in sandwich tins see above: lemon curd: lemon-flavoured butter cream see page 188: chopped toasted nuts or toasted coconut

Split the cold sponge into four layers and sandwich with lemon curd.

Ice top and sides with lemon butter cream and coat the sides with the nuts or coconut.

Pipe more butter cream on the top for decoration.

To Freeze
Freeze uncovered and then put in a box or polythene bag. Do not put anything heavy on top.

To Use
Loosen the wrapper so that it does not touch the icing. Thaw at room temperature.

RUM PUNCH GÂTEAU

Genoese sponge baked in sandwich tins see page 185: Rum: apricot jam: chocolate butter cream or butter icing: blanched almonds: crystallised violets

When the sponge is quite cold split each half in two to give four layers. Sprinkle each layer with a few drops of rum.

Soften the apricot jam if necessary with a little hot water. Flavour it with rum and sandwich the layers of cake with this.

Cover the top and sides with a layer of icing and decorate with almonds and violets.

To Freeze
Freeze uncovered and then cover with freezer paper or put in a polythene bag. Do not put anything heavy on top of the cake.

To Use
Thaw at room temperature, loosening the wrapper so that it does not touch the icing.

Icings

BUTTER CREAM
To use as a cake filling or for icing and piping. It can also be used in place of whipped cream and freezes very well. If you make

187

more than is needed for immediate use the surplus can be frozen for later use.

6 oz butter (180 g): 12 oz icing sugar (2½ c or 360 g)

Let the butter come to room temperature or put it in a warm, not hot, basin. Cream butter and sugar together.

2 egg yolks

Beat into the creamed mixture and beat until smooth and light.

Flavouring to taste

Add the flavouring. This may be a little vanilla or other essence, 2 tsp rum or liqueur, grated orange or lemon rind, 1–2 Tbs soluble coffee, 4 oz melted chocolate.

BUTTER CREAM (without egg)
Make as above but use equal weights of butter and icing sugar and omit the egg yolks.

BUTTER ICING
QUANTITIES for the inside and top of a 7 in (18 cm) sandwich

4 oz butter (1 c or 120 g): 8 oz icing sugar (1½ c or 240 g)

Soften the butter and sieve the icing sugar. Beat together until light and smooth.

Milk: water or fruit juice

Beat in enough liquid to give the right consistency for spreading or piping, thicker for piping.

CHOCOLATE BUTTER ICING
4 oz melted chocolate (120 g)

Add to the softened butter.

COFFEE BUTTER ICING
2 Tbs instant coffee powder: vanilla essence

Add with icing sugar and flavour to taste.

LIQUEUR ICING
Flavour with any liqueur, rum, brandy or sherry.

ORANGE OR LEMON BUTTER ICING

1 orange or lemon

Add the finely grated rind and use the juice for the mixing liquid.

GLACÉ ICING

8 oz icing sugar (1½ c or 240 g): about 2 Tbs hot water or fruit juice: ½ oz butter or margarine (1 Tbs or 15 g): flavouring and colouring to taste

Sieve the icing sugar into a basin. Soften the butter or margarine in half the water and add this gradually to the icing sugar, beating in well. Add more water until the icing will coat the back of a spoon without running off too freely. Add flavouring and colouring to taste. Beat well to give it a good shine.

For a large cake pour the icing in the middle and let it run over the top. If the sides are to be iced as well, let the icing run down sides and smooth evenly with a knife dipped in hot water.

For small cakes and biscuits either dip the cake or biscuit top in the icing or put it on with a small teaspoon, allowing it to fall from the tip of the spoon and run into a round.

CHOCOLATE GLACÉ ICING
Add 2 oz melted chocolate (60 g), or 4–6 Tbs cocoa powder mixed to a paste with boiling water.

COFFEE GLACÉ ICING
Add 2 Tbs instant coffee powder or to taste.

BISCUITS
As most biscuits can be kept satisfactorily in an airtight tin there is not much point in freezing them unless you want to keep them for a very long time.

Freshly baked biscuits are best of all and if a supply of raw biscuit dough is kept in the freezer a few biscuits can be baked in a very short time. A freshly baked home-made biscuit is a delightful accompaniment to cold sweets such as fruit salad, fruit fools etc.

To Freeze

Pack cooked biscuits in containers (tins are suitable if airtight) putting paper between layers.

For uncooked biscuits use recipes with a high fat content, for example see below.

Shape the dough into a roll, chill and slice. Pack the slices with Cellophane paper between each slice, reshape into a roll and freeze.

Alternatively freeze the roll without slicing and then thaw just enough to permit slicing.

Freeze piped biscuits on trays, remove and pack in layers with Cellophane paper.

To Use

Thaw cooked biscuits in unopened containers for 2–3 hours, or heat for 5–10 minutes at E 200°F G "Low". Raw ones, bake for 10–12 minutes at E 350°F G 3–4.

CLIFFORD TEA COOKIES

COOKING TIME: 12–15 mins: QUANTITIES for 5–6 doz biscuits

TEMPERATURE: E 350°F G 3–4

4 oz butter or margarine (8 Tbs or 120 g): 8 oz fine brown sugar (1 c or 240 g)

Put the fat in a mixing bowl and warm it to soften but not melt the fat. Add the sugar and beat until well blended.

1 egg

Beat into the creamed mixture.

8 oz plain flour (1½ c or 240 g): ½ tsp bicarbonate of soda: ¼ tsp salt: 2 oz chopped walnuts (½ c or 60 g)

Sift the flour, soda and salt into the creamed mixture, add the nuts and mix until well blended.

Shape the mixture roughly into a sausage shape, wrap in foil and chill in the refrigerator until firm. Then roll into a sausage-shape about 1½–2 inches in diameter, either one or two sausages.

To Freeze
The biscuits may be sliced before freezing and then re-shaped into the roll with freezer paper between the biscuits. Then wrap with double foil and freeze. I find it less trouble and just as satisfactory to freeze the mixture uncut.

To Use
If uncut allow to thaw enough to be able to slice off the number of biscuits required, about ¼ inch slices.

Put on non-stick paper on trays. Leave a little room between the biscuits for spreading. Bake until brown but not quite firm. Leave to become crisp before removing from the paper.

PINWHEEL BISCUITS

COOKING TIME: 15 mins: TEMPERATURE: E 350°F G 3–4

QUANTITIES for 24 biscuits

3 oz butter or margarine (90 g): 3 oz caster sugar (6 Tbs or 90 g)

Cream together.

1 egg yolk: ½ tsp vanilla essence

Beat in.

6 oz self-raising flour (1¼ c or 180 g): pinch salt: milk

Add to the creamed mixture with about one tablespoon milk or enough to make a stiff dough. Divide into two portions, one slightly larger than the other.

1 tsp cocoa powder: 1 tsp golden syrup

Add these to the smaller piece and knead until well blended. Roll the plain piece to a rectangle about ⅛ inch (3 mm) thick. Roll the chocolate piece slightly smaller and put it on top of the other. Roll up tightly like a swiss roll.

To Freeze
Wrap closely in double foil and freeze.

To Use

Thaw just enough to be able to slice the biscuits about ⅛ inch (3 mm) thick. Place on greased baking trays and cook until the plain part is lightly coloured. Cool on a rack.

FREEZER SHORTBREAD BISCUITS

COOKING TIME: 15 mins thawed, 20 mins frozen: TEMPERATURE: E 350°F G 3–4

QUANTITIES for 3 doz biscuits

4 oz butter or margarine (120 g): 4 oz caster sugar (1 c or 120 g): 1 egg yolk: 1 Tbs marsala or sherry

Cream the fat and sugar until light, beat in the egg and wine.

6 oz plain flour (1¼ c or 180 g): pinch salt: 1 oz cornflour (3 Tbs or 30 g)

Add these to the creamed mixture, stir and then work with the hands to make a pliable mixture.

Shape into a block like a butter pat but ¾–1 inch thick (2–2½ cm).

To Freeze

The mixture may be sliced before freezing and the pieces separated with freezer paper or freeze in the block. Wrap closely with freezer paper, seal and freeze.

To Use

Remove the number of biscuits required or thaw the block enough to be able to slice in about ¼ inch thick slices. The remainder can be re-frozen. Put a sheet of non-stick paper on the baking tray and put the biscuits on this, leaving a little room for spreading. Bake until pale brown and crisp.

Chapter Nineteen

SANDWICHES AND SAVOURIES

SANDWICHES

There are many occasions when it is convenient to be able to prepare sandwiches several days in advance. When preparing for a party is an obvious one, at holiday time, for picnics or if you have to provide daily packed lunches.

It is a good plan to have a sandwich-making session preparing several different kinds to build up a stock to last for a couple of weeks. Longer storage than this will probably mean loss of quality.

Making and Freezing

To avoid soggy results when the sandwiches are thawed it is advisable to use day-old bread, to butter liberally to prevent the filling from soaking into the bread, and to avoid putting the sandwiches in contact with the freezer plate or shelf.

If the sandwiches are to be stored for only a day or two double layers of waxed paper are suitable for wrapping. For longer periods use a freezer wrap. It is better not to wrap more than six or eight sandwiches together. If you want to make up a packet of mixed fillings, wrap each flavour separately and then over-wrap the assortment.

To Use

Thaw sandwiches in the wrapping.

If they are required for lunch they should be removed from the freezer at breakfast time and they will be thawed by the time they are needed. If the package of sandwiches is put in a container with fresh lettuce, tomatoes or other salad vegetables, the frozen sandwiches will help to keep these fresh and crisp.

To thaw in the refrigerator allow 8 hours or more; at room temperature about 4 hours.

SUITABLE FILLINGS

Avoid hard boiled eggs, salad vegetables, mayonnaise, jam, or any soggy filling.

The following give very satisfactory results:

Cottage cheese or cream cheese plus herbs and other flavourings, e.g. curry, paprika, chopped gherkins.

Other cheese.

Smoked, canned or fresh salmon.

Canned tuna fish.

Sardines.

Liver pâté.

193

Minced cold meats and pickle.

Slices of cooked ham, tongue, lamb or beef.

Asparagus rolls.

Cottage cheese with chopped anchovy fillets and mustard.

Danish blue cheese mashed with chopped ham and a little softened butter.

Spread the bread with flavoured butters, see pages 61–3, then add slices of cooked meat, or minced cooked meat.

Party Sandwiches

PINWHEELS

Use an unsliced sandwich loaf. Remove the crusts. Cut the loaf in thin slices longways so that you have slices as long as the loaf. Place each slice on a clean damp cloth. Spread generously with butter and a filling of a contrasting colour. Roll up like a Swiss roll. Wrap firmly in foil, over-wrap and freeze.

To Use

Half thaw and then cut in slices across, like a Swiss roll. Arrange on a serving plate, cover and finish thawing.

RIBBON SANDWICHES

Use sliced bread of two different colours, two white and one brown or rye; or use more brown than white. Make the sandwiches of three slices each, with the centre slice a different colour from the outsides. Use two different coloured fillings with flavours that harmonise. Wrap in foil, over-wrap and freeze.

To Use

Half thaw and then cut in fingers. Arrange on a serving dish, cover and finish thawing.

SANDWICH CAKES, TORTES OR LOAVES

For cakes use a round, flat loaf like a bap. Slice in two or three layers and fill with different flavours. The top may be iced with a savoury butter or cream cheese mixture and decorated to taste. Freeze unwrapped, then wrap and store. For a loaf use a sandwich loaf, trim off the crusts and slice it lengthwise three or four times. Sandwich the slices with fillings of different colours. Wrap very firmly and freeze. If a decorated loaf is required, after filling leave it with a board and weight on top until the layers are well pressed together, then decorate as for the cake. Freeze unwrapped then wrap and store.

To Use

For decorated ones, loosen the wrapper for thawing. For all types half thaw, cut in slices and arrange on a dish.

Cover and finish thawing.

OPEN SANDWICHES

These may be frozen provided they are made with suitable ingredients for freezing. Freeze on trays and then pack in boxes in layers with Cellophane paper between the layers.

To Use

Arrange on serving dishes, cover and thaw.

QUANTITIES FOR PREPARING A NUMBER OF SANDWICHES

One large loaf (1¾ lb or 750 g) gives about 20–30 slices.
One quartern loaf gives about 50 slices.
Allow not less than 4 oz softened butter per large loaf.
Allow 10–12 oz filling per large loaf.

SAVOURIES

Many different items suitable for savouries and snacks can be frozen either completely finished or the ingredients to make them can be frozen for quick assembly when required.

Suitable recipes for some of these will be found on the following pages:

Savoury Butters for canapés, pages 61–3.
Kipper Pasties, page 81.
Kipper Pâté, page 82.
Sardine Rolls, page 84.
Potted Duck, page 118.
Welsh Rarebit, see below.

WELSH RAREBIT

COOKING TIME: 5–10 mins; TEMPERATURE: E 400°F G 6

QUANTITIES for 8

8 slices bread: butter for spreading

Toast the bread. Trim off crusts. Let it get cold and then spread it generously with butter right to the edges.

8 oz (240 g) well flavoured cheddar or cheshire cheese: 1 Tbs cornflour: 1 tsp made mustard: salt and pepper: 5 Tbs milk

Grate the cheese and mix with the other ingredients. Be sparing with the pepper. Spread the paste over the toast.

To Freeze
Freeze whole slices or cut in pieces. Pack in layers in a box or bag with Cellophane paper between the layers. Seal and freeze.

To Use
Place in a single layer on baking trays and heat in the oven at E 400°F G 6 for 5–10 minutes or until hot. If necessary put under the grill to brown.

ALTERNATIVE
1. Freeze the mixture in portions in foil parcels packed in a polythene bag. Thaw enough to spread.
2. Keep a stock of grated cheese in the freezer for quick preparation of the recipe.

Chapter Twenty

MISCELLANEOUS

COFFEE AND TEA
If you want to buy stocks of these to last for several weeks or months they will keep fresher in the freezer. Leave them in the original packing but over-wrap with freezer paper or put in polythene bags.

ICE FOR DRINKS

Crushed ice or ice cubes can be stored in plastic bags or boxes. This is specially useful for entertaining when you will want more ice than usual, also when the refrigerator is being defrosted and you want to make sure you have some ice to hand.

It is also a good idea to have a small reserve of ice in the freezer at all times.

CATERING FOR PARTIES

Any of the following items can be purchased or made in advance and stored in the freezer:

Bread rolls, and sliced bread for toast, see page 177.

Soft cheeses, see page 67.

Hors d'Oeuvre, see page 48. Many of these items are suitable for a buffet party or in small sizes for a cocktail party.

Sandwiches, see page 192 either conventional ones or decorative pinwheels, ribbon sandwiches, sandwich cakes or loaves.

Cold Meats, see page 93.

Meat Pies and Patties, see page 169.

For a hot meat course, suitable for fork service, Goulash, see page 100; Curry, see page 97, Blanquette de Veau, see page 96, Duck in Orange Sauce, see page 117.

Cold Chicken, see page 109.

Sardine Rolls, see page 84; Kipper Pasties, see page 81.

Sweets, Ices, see page 148; Cheese Cake, see page 159.

Frozen Chocolate Cake, see page 161; Bavarois, see page 158.

Apple Snow, see page 157; Fruit Mousses, see page 163.

Chantilly Cream to serve with fruit salad and fresh fruits such as strawberries and raspberries, see page 64.

Freezer Shortbread and other Biscuits, see page 192.

Continental Apple Flan, see page 173.

Sponge Gâteaux and other Gâteaux, page 185.

SPECIAL DIETS

A freezer makes catering for special diets much less of a burden than it might be otherwise. Enough of a special food can be made for several meals and the surplus frozen. This avoids cooking single portions frequently, or the more common monotonous diet due to having to use up the food before it goes bad.

Suggestions for Types of Food to be Frozen

PURÉES

These can be of fruit, vegetables, meat, fish or any other food that has to be sieved.

Cook, sieve, cool quickly by standing the container in iced water. Pour into small boxes, cartons or polythene bags, one portion size. Seal and freeze.

To Use

Thaw in the refrigerator or at room temperature for using cold, heat in a double boiler for serving hot, or add to hot sauces or other liquids and allow to melt and heat. They are useful for infant feeding, low fibre diets, soft or semi-solid diets and for people with sore mouths and throats.

MINCED FOODS

Many diets allow certain meats only if they are minced. These may be frozen raw or cooked, the latter with gravy or sauce and ready for heating and serving. Freeze in small amounts for one portion.

GLUTEN-FREE FOODS

Bread, sponges, cakes, biscuits, puddings and sweets made for the gluten-free diet freeze very well. Savoury dishes containing gluten-free sauces or the sauces by themselves are other useful items.

LOW-SALT FOODS

The most useful items for this diet are the salt-free bread and special savoury dishes.

LOW CALORIE AND LOW CARBOHYDRATE DIETS

Especially useful for these are fresh fruit frozen without sugar, purées without sugar, unsweetened fruit juices and low-calorie sweets and savoury dishes.

LOW-FAT DIETS
Freeze sauces, sweets and savoury dishes suitable for this diet.

PLATED DINNERS
These are useful meals for the family when you are away from home.

Use any meat suitable for freezing. It should be cooked and covered with gravy or a sauce.

Cooked rice or mashed potatoes are suitable.

Any other vegetables should be prepared and blanched as for freezing.

To Freeze
Use the special foil compartment plates made for this purpose. Put the meat and vegetables in separate compartments. Wrap in foil and freeze.

To Use
Put in the oven, still covered with foil, at E 375°F G 5 for 30–45 mins or until thoroughly hot.

INDEX

BEST SELLING MAYFLOWER TITLES

☐	583 11692 2	THE MOVIE MAKER	Herbert Kastle	8/-
☐	583 11574 8	THE "F" CERTIFICATE	David Gurney	7/-
☐	583 11717 1	LAST SUMMER	Evan Hunter	5/-
☐	583 11683 3	FANNY HILL (unexpurgated)	John Cleland	7/-
☐	583 11186 6	WANDERERS EASTWARD, WANDERERS WEST	Kathleen Winsor	8/-
☐	583 11650 7	GOAT SONG	Frank Yerby	8/-
☐	583 11710 4	THE DEAL	William Marshall	8/-
☐	583 11715 5	THE MADONNA COMPLEX	Norman Bogner	7/-
☐	583 11734 1	WORLD ATHLETICS & TRACK EVENTS HANDBOOK/70	Compiled by Bruce Tulloh	5/-
☐	583 11703 1	WORLD MOTOR RACING & RALLYING HANDBOOK/70	Compiled by Mark Kahn	5/-
☐	583 11711 2	OCTOPUS HILL	Stanley Morgan	5/-
☐	583 11377 X	THE SEWING MACHINE MAN	Stanley Morgan	5/-
☐	583 11336 2	STILETTO	Harold Robbins	5/-
☐	583 10339 1	ETERNAL FIRE	Calder Willingham	7/-
☐	583 10429 0	LUCY CROWN	Irwin Shaw	7/-
☐	583 11665 5	THE COUNTRY TEAM	Robin Moore	8/-
☐	583 11342 7	DYNASTY OF DEATH	Taylor Caldwell	8/-
☐	583 11530 6	THE BEATLES	Hunter Davies	8/-
☐	583 10437 1	THE MARRIAGE ART	J E Eichenlaub	5/-
☐	583 11563 2	OR I'LL DRESS YOU IN MOURNING	Larry Collins & Dominique Lapierre	7/-
☐	583 11443 1	THE MAN WHO HAD POWER OVER WOMEN	Gordon M Williams	6/-
☐	583 11157 2	THE CAMP	Gordon M Williams	5/-
☐	583 11608 6	LOVE IS A WELL-RAPED WORD	Doreen Wayne	6/-
☐	583 11600 0	THE BUSINESS OF MURDER	Edgar Lustgarten	5/-
☐	583 11338 9	THE JET SET	Burton Wohl	6/-

All these books are available at your local bookshop or newsagent; or can be ordered direct from the publisher. Just tick the titles you want and fill in the form below.

Write to Barnicote, P.O. Box 11, Falmouth. Please send cheque or postal order value of the cover price plus 9d. for postage and packing.

Name...

Address..

....